The Product Leader

Product Management Leadership
by Example

The Product Leader

Product Management Leadership by Example

Joseph P. Lee

thieros.
press

Published by Thieros Press, an imprint of Thieros Corporation, Chicago, Illinois, USA

Thieros Press provides this book in printed, electronic, and audio format. Please check with your favorite online retailer, local retailer, or book reseller for details.

Thieros.com

The Library of Congress Cataloging-in-Publication Data (LCCN) is available upon request.

ISBN 9798998582745 (Hardcover)
ISBN 9798998582752 (Paperback)
ISBN 9798998582769 (eBook)
ISBN 9798998582776 (Audiobook)

This book was written and designed by a human, using Serif's Affinity Publisher 2.

This book is dedicated to my parents, Shirley and Ed. They sneaked my first computer under the Christmas tree when I was 16, even though we were far from wealthy. It was a Texas Instruments TI-99/4A, and it was a real doozy!

For the next year, I spent all my free time locked in my bedroom closet with a tiny desk, stool, and my new computer. It was quite a setup, enabling me to spew out countless, useless programs which I thought would revolutionize the world.

That one gift ignited a madness never tamed and often reckless.

Who This Book is For

This book is a must-read for anyone involved in product development, whether you're a seasoned pro or just starting out. It's ideal for those dreaming of becoming a leader or looking to take their skills to the next level. Even if you're new to product management or don't have a traditional role, this book can help you get started.

What this book is *not* is a manual for every little thing in a product manager's daily life. It is written for future and existing product leaders. I hope it will help you succeed in interviews for that next promotion, give you more confidence, and set you up for success.

I do my best to write in simple, straightforward language, using few jargon terms. Whenever I come across an acronym that I hear a million times a day, I don't assume you know its meaning and I'll define it. And if there's a new technology or term that you might not be familiar with, I'll do my best to introduce you to it.

While not all product managers and leaders are in the technical space, a lot of what I write about relates to applications and digital technologies (Internet, mobile apps, etc.). This does not mean the skills learned are not transferable to other types of business products. In fact, becoming a great product leader trains you to adapt to all kinds of industries and products. You will find that the concepts I write about can be easily adapted to your quest, as well.

How This Book Was Organized & Written

I see product leadership in two ways: Joey, the leader, and the thing Joey is currently working on. In other words, **The Role** and **The Product**. So, this book is divided into those two sections.

In the first section, we'll explore the product organization, your job as a product leader, how to work well with others, some essential soft skills to build, and how great teams are structured in well-organized businesses.

The second section, focuses on building solid products, growing them responsibly, learning how to increase revenue, and making them valuable and loved by your customers.

I write short examples after many topics and always at the end of each chapter. Thus, "by Example" became an essential part of this book's subtitle. The names of companies and individuals contained within examples have been changed, in order to protect the innocent.

Lastly, this book was authored by a human. The human author selected the topics, composed the content, meticulously edited the text, constructed the index, and designed and created the layout. Artificial intelligence (AI) was employed only to enhance the quality of certain paragraphs, particularly when the human author experienced excessive excitement or when his thoughts were unacceptably disorganized.

Contents

About Me   xiii

Introduction   xv

Part 1 – The Role

Chapter 1 – Building a Product-Centric Organization   3

Defining & Adopting Product Centricity   3

When Product Centricity Fails   7

Architecture of Product Organizations   12

Choosing a Product Architecture   14

Concepts in Action: FlowSimple   16

Execution Essentials   18

Chapter 2 – Industry Established Leadership Roles   21

The Ripple Effects of Misalignment   22

Benefits of Proper Title Structure   23

Senior Product Manager   27

Group Product Manager   33

Director of Product Management   39

Vice President of Product   45

Chief Product Officer   53

Concepts in Action: Brovis Technologies   59

Execution Essentials   62

Chapter 3 – Specialized Product Roles   65

Looking Ahead   65

Technical Product Manager (TPM)   67

Growth Product Manager   69

Platform Product Manager   71

AI/ML Product Manager   73

Data Product Manager   75

Concepts in Action: 30Fusion   77

Execution Essentials   80

Chapter 4 – Managing Great Teams   83

Lane Keeping & Team Hierarchy   83

Creating a Culture of Trust   89

Onboarding for Future Success   92
Polishing the Team   95
Concepts in Action: Sorensen Technologies   104
Execution Essentials   107

Chapter 5 – Stakeholder Management   109

Defining Stakeholders   109
Building Stakeholder Relationships   111
Stakeholder Communication   115
Potential Stakeholder Pitfalls   120
Stakeholder Management Tools   122
Conclusion   124
Concepts in Action: CloudTwenty   125
Execution Essentials   127

Chapter 6 – Leading Cross-Functional Teams   129

Objectives, Goals & Indicators   130
Meetings & Communication   132
Keeping Connections   138
Conclusion   140
Concepts in Action: CargoTrac Technologies   141
Execution Essentials   145

Chapter 7 – On-site, Hybrid & Remote Teams   149

Promote Flexibility & Maturity   149
Maintaining Strong Communication   152
Requiring Discipline and Accountability   156
Conclusion   158
Concepts in Action: Noviki Brass   159
Execution Essentials   163

Chapter 8 – Developing Future Leaders   165

Coaching   166
Mentoring   168
Servant Leadership   171
Conclusion   174
Concepts in Action: ElevationTech   176
Execution Essentials   179

Chapter 9 – Sustainability & Ethical Leadership   181

Sustainability   181
Ethical Leadership in Product   187
Conclusion   192
Concepts in Action: InnoEco Solutions   194
Execution Essentials   196

Part 2 – The Product

Chapter 10 – Product Alignment & Business Strategy   201

Strategic Vision   201
Market & Customer Alignment   204
Business Performance Metrics   209
Organizational Synergy   212
Execution & Operation   216
Conclusion   218
Bonus Tip   220
Concepts in Action: PixelRod Software   221
Execution Essentials   225

Chapter 11 – Competitive Analysis   227

Market Positioning & Differentiation   228
Competitive Intelligence Gathering   231
Strategic Response Planning   235
Continuous Monitoring & Analysis   237
Conclusion   240
Concepts in Action: BrewChoice   241
Execution Essentials   245

Chapter 12 – Product-Led Growth   247

Product-Market Fit   248
The Four Pillars of Sustainable Growth   248
Measuring What Matters   252
Building a Growth-Oriented Organization   254
Growth Pitfalls to Avoid   256
The Future of Product Growth   258
Conclusion   259
Concepts in Action: FlavorVault   260
Execution Essentials   265

Chapter 13 – Smart Monetization   267

The Psychology of Value Perception   267
Designing Your Monetization Model   269
Executing Your Monetization Strategy   271
Monetization Maturity   274
Monetization Pitfalls to Avoid   275
Conclusion   277
Concepts in Action: Zigler   278
Execution Essentials   281

Chapter 14 – Mastering Data for Decision Making   283

Data Collection & Quality   283
Analysis Methods   286
Metrics That Matter   289
Decision Architecture   292
Data Communication   294
Conclusion   298
Concepts in Action: MariTrac   299
Execution Essentials   301

Chapter 15 – Extensibility, Partnerships & Integrations   303

Product Extensibility   304
Strategic Partnerships   309
Platform Integration   313
Conclusion   316
Concepts in Action: TractorTech   318
Execution Essentials   321

Chapter 16 – Go-To-Market Planning   323

Channels & Distribution   324
Marketing & Demand Generation   330
Post-Launch   335
Conclusion   339
Concepts in Action: Orchard   340
Execution Essentials   343

Index   **347**

About Me

I'd like to share my background and why I'm qualified to write this book. My career path may have similarities to yours, though I've had a somewhat broader journey than most, working across many different products.

I began as a software engineer and became one of the most sought-after vendors at Microsoft during its early growth period. I've founded multiple companies, primarily in the Internet and software industries.

I enjoy building all types of products, with particular interest in digital innovation: websites, mobile apps, native applications, APIs, and developer tools. I believe in servant leadership and find fulfillment in the success of others.

Product management found me rather than the other way around. I was working in this space before it was commonly known as "product management."

My engineering background gave me the technical understanding to comprehend my team's capabilities and the technologies needed for effective solutions. I naturally became the person who communicated our capabilities to the business side while helping define deliverables, identify customer needs, and develop products.

As I gained experience, my responsibilities expanded to include customer meetings, initiative support, feature prioritization, and go-to-market assistance. Eventually, this role became known as a "product manager" — someone who represents the customer while coordinating team members across the organization, like a hub connecting spokes spanning the enterprise.

I spent over eight years at Microsoft working on numerous internal and retail products as a prime vendor. After selling my small firm to a larger company (later absorbed by ManpowerGroup), I started TangoWire.

TangoWire was a web application service provider (now known as SaaS) for online communities, their owners, and marketing partners. This double-sided marketplace required diverse technical and product management skills. My Microsoft experience proved invaluable at that time. After nearly nine years, I sold my majority shares in 2007.

I discovered that product management — understanding customer needs and transforming them into actual products — could be both rewarding and profitable. This led me to become a serial entrepreneur.

My career also continued through roles at large banks and consulting for various organizations of different sizes. I usually maintain a clientele of individuals seeking product career coaching.

This book draws from over thirty years of experience. My path hasn't been linear — I still move between engineering and business ventures as interests dictate. Throughout it all, being a product manager has remained my professional core.

Introduction

Throughout my career as a product leader, I've watched firsthand the extraordinary evolution of product management. What began as a specialized function has become one of the most sought-after roles across nearly all industries. You're likely somewhere on this journey yourself. Maybe you're transitioning from being an individual contributor to your first leadership position. Or maybe you're already leading teams and needing to refine your approach.

The Rise of Product Management

Product management has experienced some crazy growth over the past decade. Companies have found that building successful products requires dedicated professionals who understand both business strategy and their customers' needs. The demand for skilled product managers continues to increase year after year. I remember when explaining my job title meant a ten-minute conversation. Now, product management has become a well-understood and respected field.

The Maturing Practice

As our field has grown, so has its sophistication. Product management is no longer simply about gathering requirements or managing issue backlogs. Today's product organizations use complex methods, advanced data analytics, and operate within well defined organizational structures. More importantly, I'm seeing a generation of seasoned product managers ready to step into leadership roles. This evolution creates both opportunities and challenges that this book addresses directly.

Your Pathway to Executive Leadership

One particularly exciting aspect of product leadership is its potential for being a pathway into executive roles. Product leaders develop a unique combination of skills: strategic thinking, cross-functional collaboration, customer empathy, technical understanding, and business acumen. These skills prepare you for higher organizational leadership. Many of today's CEOs and executives began their careers in product management. This book will help you build the foundation for that trajectory should you choose to pursue it.

The following chapters provide practical guidance for your journey into product leadership. I begin by defining the various product leadership roles that exist today; all the way up to Chief Product Officer. Understanding these roles will help you identify your next career move.

Mastering Stakeholder Management

Success as a product leader hinges on your ability to effectively manage stakeholders across the organization. I'll explore techniques for building relationships with critical collaborators. I've seen even brilliant product strategies fail due to poor stakeholder management, and I'll share some tactics that have consistently worked throughout my career.

Embracing Servant Leadership

The most effective product leaders understand that their primary duty is to help their teams to succeed. I'll cover how to mentor product managers, remove obstacles, and create environments where product thinking can flourish. I'll share specific techniques for supporting your team while still providing clear direction and promoting accountability.

Achieving Product Alignment

As organizations grow, maintaining alignment becomes increasingly difficult. I will examine how to create shared understanding across teams, establish clear decision-making processes, and develop a way of measuring performance that keeps everyone moving in the same direction.

Cultivating Product-Led Growth

I'll also explore how product leadership extends beyond the product organization to influence company strategy. You'll learn how to foster a product-led mentality throughout your organization, connecting product decisions to business outcomes while maintaining customer-centricity.

Your Leadership Journey Begins

This book represents the guidance I wish I'd had when stepping into my first leadership role. The path ahead contains challenges, of course, but it also contains tremendous opportunities to shape products that impact thousands or even millions of lives. My goal is to provide you with practical wisdom that helps you grow as a leader.

The product management world is constantly changing, and so are the roles of product leadership. By learning the skills in this book, you'll be ready to handle today's challenges and be prepared for the future of our profession.

There's lots to cover. Let's go!

Part 1 –

The Role

The Role

Chapter 1

Building a Product-Centric Organization

First, let me state an important truth: no organization has a typical setup. The roles, responsibilities, and reporting structures of product teams vary wildly across companies. What works for a 50-person startup may not work for a 50,000-person enterprise. What functions smoothly in a B2B SaaS company might crash and burn in a consumer tech firm. Consider this chapter a map with the general terrain marked, not turn-by-turn GPS directions for your specific journey.

It is a product leader's responsibility to bring product centricity to their organization and it's not an easy task. However, it can be done and this first chapter focuses on the type of organizations where leaders thrive. We must start here, or else the industry established roles in the next chapter will be setup for failure. Yes, I'm throwing you into a big topic, right out of the gate and headfirst. Let's go!

Defining & Adopting Product Centricity

Product centricity is the radical notion that your product should actually solve problems for your customers. Revolutionary, I know. It means placing the product experience at the heart of your business strategy rather than treating it as an afterthought that marketing needs to compensate for with clever ads featuring talking animals.

In a product-centric organization, every decision begins with the question: "How does this improve our product's ability to solve our customers' problems?" This stands in stark contrast to sales-led organizations where the primary question is "Can we sell this?" or marketing-led organizations where it's "Can we convince people they need this?"

A truly product-led company has several distinct characteristics that set it apart from its less enlightened cousins.

Customer Obsession

Product-led companies don't just listen to customers. They observe them, study them, and figure out their needs sometimes better than customers themselves. It's like being a detective for user problems and your tools of the trade are data and feedback.

True customer obsession is like being an anthropologist; studying an unfamiliar tribe. You're not just asking customers what they want; you're embedding yourself in their natural habitat to observe behaviors and habits they don't even know they have.

At Microsoft, I once spent three days shadowing the pre-sales technical team — not exactly a glamorous job, but it revealed huge inefficiencies our initial overview never captured. Problems coordinating through email tools and broken CRM workflows became the story for our product roadmap for the coming year. The team never articulated these pain points because they'd normalized their painful workflows, like how you don't notice the buzz of a refrigerator until it stops.

In truly customer-obsessed companies, knowledge about customers isn't hoarded by research teams or executives. Engineering teams pay careful attention to user sessions and often listen-in on support calls. Executives regularly speak with actual customers who aren't just the friendly ones cherry-picked for board meetings.

I've seen companies transform when product leaders and developers start periodically spending time watching customers and end users struggle with features they built. Code quality improves when developers suddenly see the angry faces of customers using their features. The bug that seemed minor in the backlog becomes urgent when you've watched someone scream while tripping over it.

Customer obsession ultimately means building company-wide empathy — a shared, visceral understanding of your users' world that makes their problems your problems. When that happens, product decisions stop being abstract exercises and start feeling personal. And that's when the magic happens.

Decision-Making Through the Product Lens

Traditional companies operate with a top-down decision model. The CEO finds a shiny penny, executives write feature demands, and product teams implement without question, occasionally staging small rebellions that are swiftly crushed.

Product-led companies flip this hierarchy on its head. Product insights bubble up from customer interactions and data, forming the foundation upon which strategic decisions are built. Executive decisions aren't arbitrary mandates but rather confirmations of patterns already evident in product data.

Product-centric decision-making applies what I call the "So What?" test to every potential initiative. It's not enough for something to be technically impressive or for competitors to have it. The critical question becomes: "So what? How does this materially improve our customers' experience or outcomes?"

This simple filter cuts through the feature bloat that plagues so many products. I've seen roadmap meetings transform when every proposed feature faces this question. Suddenly, the engineering team's desire to rewrite the backend in the latest method becomes less compelling when it can't answer the "So What?" test from a customer perspective.

Product Centered, Cross-Functional Collaboration

In most organizations, cross-functional collaboration just doesn't seem to work well. In meetings, everyone seems to speak different languages, nod politely, and leave with entirely different interpretations of what just happened. Product-led companies, however, transform such a catastrophe into a cohesive melody by establishing the product as a universal language.

Traditional companies operate in team silos that feel like territorial fiefdoms. Marketing creates campaigns without understanding the company's limitations. Engineering builds features without understanding customer usage contexts. Sales promises features that were drawn in PowerPoint by one of their team members.

Product-centric organizations tear down these barriers by centering every function around a shared product vision. The product becomes the center of everything they do, meet about, share, and celebrate.

I witnessed a large company transform into product centricity, when it once couldn't even get through a quarterly planning meeting without teams becoming disgruntled. By establishing the product roadmap as the central planning document — informed by customer data and mapped to business outcomes — the entire dynamic shifted. Suddenly, marketing wasn't planning campaigns in isolation but synchronizing them with feature releases. Sales stopped selling vapor and started highlighting real product features. The shared product focus created a whole new playing field and the customer felt it.

Effective cross-functional collaboration doesn't mean endless meetings that drain everyone's will to live. On the contrary; it requires structured processes for input, prioritization, and decision-making.

Many successful product-led companies implement some version of a product council; a cross-functional team that regularly reviews product performance, prioritizes initiatives, and aligns department efforts. Unlike traditional committees that debate endlessly, these councils operate with clear decision processes based on product impact.

Example: MexiSprint

MexiSprint, a food delivery app in the Southwest U.S., was struggling with team alignment. Their marketing team would create promotions for geographic areas the team couldn't reliably get to. Their engineering team built features delivery drivers found confusing. Customer support was constantly blindsided by app updates.

CEO Yolanda decided to make a change. She established the product roadmap as their central planning document and created a "product council" with representatives from each department that met weekly.

Instead of departments working in isolation, everything now revolved around the product roadmap, which was created using actual customer data and tied to specific business outcomes. Marketing planned promotions around reliable destinations. Engineering built features that drivers actually needed. Support could properly prepare for upcoming changes.

The best part? Meetings became more productive. When the VP of Sales wanted to add a feature for one big restaurant chain, the council evaluated it against their agreed-upon "So What?" test; would it actually improve the experience for most customers? Sometimes the answer was yes, sometimes no, but decisions were made based on product impact rather than whoever argued loudest.

Six months later, their delivery completion rates jumped 30%. The app became more focused, with fewer confusing features that nobody used (or wanted). Teams worked together with a shared purpose, and the customers could taste the difference in their delicious, on-time tacos.

Metrics That Matter

In the world of measuring product success, we've created an odd phenomenon; companies drowning in data while starving for insight. The typical executive dashboard is far too complicated, with dozens of charts and indicators that create the illusion of control and which look like an airplane cockpit. None of this is insight.

Traditional companies gorge themselves on vanity metrics — those impressive-sounding numbers that inflate egos but predict nothing about the product or business health. Downloads, registered users, page views, and social media mentions fall into this category. They're the corporate equivalent of counting Instagram likes. They might make you feel popular, but they won't pay the bills.

Product-led companies ruthlessly distinguish between metrics that look good in slide decks and metrics that actually predict success. They know that 100 devoted users create substantially more value than 10,000 drive-by visitors.

Product-led organizations find their North Star metric; the single measurement that best captures the core value their product delivers to customers. This isn't a financial metric but rather a usage metric that indicates customers are getting something from the product.

For Spotify, it's "time spent listening." For Airbnb, it's "nights booked." For Slack, it's "messages sent between teams." These metrics perfectly capture product value in a way revenue alone never could.

The power of a well-chosen North Star metric is its ability to align cross-functional teams. When everyone from engineering to marketing to customer support understands what single needle they're trying to move, decision-making becomes universally clear. Trade-offs that once caused endless debate suddenly have obvious answers when viewed through the lens of your North Star.

Within this book, there is a chapter dedicated to mastering data for better decision making.

The Payoff of Product Centricity

Organizations that embrace product centricity enjoy faster growth, higher retention rates, more efficient customer acquisition, and ultimately, sustainable competitive advantage. When your product sells itself, your marginal cost of customer acquisition drops dramatically, and you can invest those savings into making your product even better — a virtuous cycle that leaves competition scrambling to keep up.

The truth is, in today's market, product centricity isn't just a nice-to-have philosophy; it's a survival strategy. Because when customers have unlimited choices and can switch with a click, your product needs to be more than just functional. It needs to be indispensable.

And that, my friends, is a product-led, product-centric company.

When Product Centricity Fails

I've spent many years listening to companies declare themselves "product-centric" with great enthusiasm. The transformation typically begins with an executive retreat where everyone wears their Friday clothes and uses phrases like "customer-obsessed" and "product-led growth." There's usually an inspirational talk about Spotify's squad model or how Amazon writes press releases for products that don't yet exist. Everyone leaves with colorful sticky notes, a new subscription to Miro or Figma, and a feeling they can tackle a hole new onslaught of features. Hooray!

Six months later, the same executives are wondering why they still can't ship features faster than their competitors.

Becoming truly product-centric requires more than renaming your "Project Managers" to "Product Managers" and asking engineers to sit with designers twice a week. It demands fundamental restructuring of how decisions are made company-wide, how success is measured, and most importantly, how power flows through roles in the organization.

The Organizational Immune System

Organizations, like bodies, have immune systems. When you introduce a foreign concept — like genuine product centricity — the corporate antibodies, with impressive efficiency, mobilize and attack. I've watched this happen in companies; from nimble startups to huge enterprises, and the pattern is remarkably consistent.

The first sign of resistance usually appears when a product decision conflicts with a sales commitment. Suddenly, the thought of "building for scalable user needs rather than one-off customer requests" evaporates quickly. The VP of Sales, who has quarterly targets tied to his bonus, makes an impassioned case about losing a strategic client. The CEO, remembering the board's focus on revenue growth, ultimately sides with Sales.

Just like that, your product roadmap has a new priority item. It happens all the damned time.

This is the moment most organizations fail their first test of product centricity. And it's not because prioritizing a specific customer request was necessarily wrong — sometimes it's the right call. The failure is in how the decision was made; through power dynamics and financial pressure rather than through the product decision process supposedly put in place during that wonderful offsite. So much for all those Figma subscriptions!

The Three Types of Failed Product Centricity

In my experience, three organizational anti-patterns consistently emerge to undermine product-centric aspirations:

The Matrix of Confusion

Most companies attempting product centricity create some form of matrix organization, with functional reporting lines intersecting with product-oriented teams. On paper, it looks elegant. In practice, it creates employees with two bosses and conflicting priorities.

I once watched a designer caught between her design director (who wanted her to maintain brand consistency) and her product manager (who wanted to test a radically different approach based on user research). The poor woman spent more time managing internal politics than designing solutions. Meanwhile, the competition shipped three new features.

The Budget Betrayal

Nothing reveals organizational priorities more honestly than budgeting processes. Companies talk about being product-led while maintaining project-based budgeting cycles where funding must be secured annually based on detailed business cases forecasting ROI with suspicious precision.

True product centricity requires stable, dedicated funding for product teams with the autonomy to pivot based on what they learn. Otherwise, you've created the organizational equivalent of telling someone to "think outside the box" while keeping them on a short leash.

The Metrics Mirage

I've sat through countless executive reviews where product teams present engagement metrics while the CFO waits impatiently to discuss revenue. In truly product-centric companies, core product metrics are treated as leading indicators for financial outcomes, not as cute distractions from "real business metrics."

When your CEO can't name your product's primary engagement metric but can recite monthly booking targets by region, you're not a product-centric company; regardless of what your career site claims.

The Painful Path to True Product Centricity

The uncomfortable truth is that becoming genuinely product-centric requires dismantling power structures that have served executives well for years. It means telling your star salesperson they don't get to dictate product features. It means allowing product teams to kill projects that aren't working, even after significant investment. It means accepting quarters where revenue might dip while you rebuild foundations.

Few leadership teams have the stomach for this transformation, which is why most settle for product centricity theater — the organizational equivalent of rearranging deck chairs on the Titanic while calling it a naval innovation strategy.

The companies that do make this transition successfully share three characteristics:

Courage at the Top

CEOs who protect product teams from quarterly pressure when appropriate, who model data-driven decision making even when the data delivers unwelcome news.

Clarity in Structure

Simplified organizational designs where authority clearly aligns with accountability, and where product decisions don't require approval from seven different stakeholders with competing incentives.

Metrics Consistency

A unified measurement method where product health indicators connect visibly to business outcomes, allowing everyone to speak the same language when discussing success.

The first casualty on the path to product centricity is often the leadership's pet projects. Nothing tests an organization's commitment to product-led decision-making like killing the CEO's favorite initiative because the data shows customers simply don't care about it.

I've witnessed boardroom scenes that could rival Shakespearean tragedies — executives confronted with indisputable evidence that their strategic vision doesn't align with customer behavior. The most telling moment isn't the presentation of the data but what happens immediately after. Do they question the methodology, blame the implementation, or actually change course? In truly product-led companies, even the most senior leaders subordinate their intuition to evidence.

Additionally, the shift to product centricity often requires abandoning comforting rituals like rigid annual roadmaps and inflexible quarterly commitments. This creates genuine organizational anxiety, particularly among executives accustomed to the illusion of predictability.

Perhaps the most excruciating transition is shifting performance metrics. When you've built your professional identity around hitting certain numbers, having those numbers suddenly declared irrelevant can trigger existential panic.

Marketing leaders who've measured success by lead volume resist being measured on qualified user activation. Sales executives compensated on new logos push back against expansion revenue metrics. Engineering managers evaluated on velocity struggle with being measured on feature adoption.

These reactions aren't irrational; they're human. People optimize for what they're measured on, and changing those metrics feels like changing the rules mid-game. The organizations that successfully navigate this transition acknowledge the discomfort rather than dismissing it.

Example: DynaClean

DynaClean started as your typical cleaning service with an app tacked on as an afterthought — a booking system that frustrated both cleaners and customers alike. Their CEO, Brenda Jacobson, had declared them "product-centric" at a beach resort retreat where everyone wore matching company polos and wrote service ideas on fancy digital whiteboards.

Months later, they were still struggling. The sales team continued adding service guarantees whenever a retirement community threatened to split for their competitor. Engineering was building app features no one used or asked for. Executives measured success by new client acquisitions rather than repeat bookings or employee retention.

The turning point came during a tense investor meeting where Brenda had to explain why their employee turnover kept climbing despite raising wages. With a little nudging from an experienced board member, she realized they were practicing product centricity theater while maintaining the same old power structures.

Their transformation began with three key changes:

- Brenda started protecting the product team from monthly revenue pressure, giving them permission to say no to one-off customer requests that complicated the employee experience.

- They simplified decision-making by establishing clear product ownership; the PM gained actual authority to make final calls on the app features based on usage data from both customers and employees.

- They unified their metrics framework so everyone spoke the same language — their North Star metric (employee satisfaction score) connected directly to customer retention and revenue outcomes.

The real test came when they killed their "premium home bug exterminating service" — Brenda's pet project she'd been cramming down everyone's throat for a year — because testing showed customers simply didn't want those treatments or chemicals in their homes. The data won, not politics.

Two quarters later, employee retention improved dramatically; customer satisfaction followed, and eventually, so did revenue. They added fewer bells and whistles but made the app actually useful for the people who mattered most — their cleaning professionals.

It wasn't all smooth sailing. Their star retirement facility sales director quit when told she couldn't promise custom cleaning features. They missed growth targets for two quarters while rebuilding foundations. Some executives struggled with their diminished influence on product decisions.

But eventually, DynaClean became genuinely product-centric — not because they claimed so in off-site happy hours, but because they fundamentally changed how power and decisions flowed through the organization.

The Light at the End of the Tunnel

This might sound depressing, like I'm suggesting true product centricity is an unachievable ideal. That's not the case. I've seen organizations transform successfully, though rarely without pain and never overnight.

The most successful transitions I've witnessed have been evolutionary rather than revolutionary. They begin with small, empowered product teams operating as proof points within traditional structures. As these teams demonstrate success — shipping faster, adapting more nimbly, and ultimately driving business results — they earn the right to operate with greater autonomy. Gradually, their models expand across the organization.

The journey to product centricity isn't about dramatic reorganizations announced in company-wide meetings. It's about the patient, persistent realignment of incentives, the careful cultivation of new capabilities, and the gradual shift in how decisions are made.

In other words, becoming product-centric is itself a product challenge; one requiring the same user-focused, iterative, and evidence-based approach we apply to our actual products. And like most significant product challenges, it's never really finished. There is no transformation end date when you can declare victory and move on.

But there is a day when you realize your organization makes decisions differently than it once did. When customer insights drive strategy rather than merely informing it. When teams move with autonomy toward clear outcomes rather than blindly executing predefined outputs.

That's the day you've become genuinely product-centric. And trust me! You'll know it when you get there.

Architecture of Product Organizations

Product organizations generally form the bridge between what's technically possible and what's commercially viable. We translate customer needs into technical requirements, business objectives into product roadmaps, and market opportunities into strategic bets.

There are three distinct product organization models I've experienced when working at large and small companies. Why and how each of them were implemented truly depended on who the executive leaders were and what their past experience was. Let's explore these models.

The Customer Journey Model

Here, product teams align with stages of the customer journey. You might have acquisition, activation, revenue, retention, and referral teams. Someone once coined this the AARRR "pirate metrics" method.

Picture a payroll processing website. One product manager might team up with marketing to lure in new clients. Another PM will take care of pricing and revenue generation. A third PM will focus on the website's functionality, customer outreach, improvements, and new features. Lastly, a product manager will oversee incentives for getting new clients from existing ones through a partner program. In this example, the product teams are divided into the different aspects of the customer experience. In this example, they represent the acquisition, revenue, retention, and referral aspects of the customer's experience (journey).

I once implemented this model at a B2C online community. Our retention team focused exclusively on reducing member churn, our activation team made sure features were understood and used, while I optimized our monetization. This approach worked well until we realized that some initiatives that improved retention hurt revenue, and vice versa. The model created natural tensions that, while healthy in moderation, sometimes devolved into territorial disputes.

I am a big fan of diagramming and disseminating customer journeys. It's important to see what's happening under the hood and where the hiccups are.

Where this model shines is cohesiveness. When a team owns the entire "onboarding journey," that experience tends to feel like it was designed by humans rather than competing committees. The challenge? Your team might master one journey segment while remaining blissfully unaware of what happens before or after their slice of the customer pie.

The Outcome-Based Model

This increasingly popular approach organizes teams around business outcomes rather than features or functions. Teams might be structured around metrics like "increase user engagement" or "improve conversion rates."

The downside? Sometimes teams get so focused on their metric that they develop tunnel vision — improving their number while accidentally torpedoing someone else's. I've seen growth teams celebrate hitting their acquisition targets while retention quietly plummeted. Awkward.

When I was a product leader at a very large bank, we attempted to transform from a feature-factory to outcome-based teams. Instead of having teams responsible for building messaging features, we created a "customer engagement team" measured by how often users interacted with our bankers. This shift was supposed to lead to a more cohesive, user-centric approach to product development and I totally get it, but it didn't exactly live up to its hype.

That said, that doesn't mean this model won't work for some teams. Focusing on the right metrics, while not ignoring holistic, product-wide outcomes, is the key.

The Functional Model

In this setup, product managers are organized by function or domain expertise. You might have separate product teams for mobile, web, platform, payments, and so forth. This can also be delineated by technologies across all these domains. For example; the security product team might focus on log in/out screens and pages across the entire technology portfolio, while the account product team focuses on account/profile screens and pages. This is similar to a customer journey model, except the focus is on the function itself, not a stage in a customer's journey.

During my time at Microsoft, this was the model which primarily prevailed. The Windows Server teams obsessed over security and networking, while the Microsoft Office teams fixated on office productivity, collaboration, and fixing Outlook defects. These teams rarely spoke to each other, which often created a disjointed user experience. Our customers didn't think in terms of "servers" or "email agents" – they just wanted a seamless experience.

I have seen this type of structure work most often. In this model, you soon end up with some serious subject matter experts (SMEs, pronounced "smees"). The team's

Pms start to become more technically knowledgeable and that's invaluable. Additionally, these folks usually stay around a lot longer because they're needed experts, and that lends itself to job satisfaction and thus dependability.

To make this model work, product managers must learn soft skills and to be solid communicators. They should be experts at using product and project tracking tools so that dependencies across the organization are monitored and addressed.

Another requirement is the existence of a global roadmap. Without a holistic understanding, by all, of what's "in the works" across the organization, teams can often develop fiefdoms and territorial secrecy – not realizing they are impacting important functional or business strategies. With a globally shared and transparent roadmap, cross-functional teams communicate better, as they better understand each other's efforts.

Choosing a Product Architecture

Selecting the right product management architecture (model) is less like following a recipe and more like jazz improvisation. There are principles to guide you, but the magic happens when you adapt to your unique circumstances. Your product's maturity, market dynamics, and team composition all play crucial roles in this decision.

I've seen brilliant product leaders stumble by implementing theoretically perfect models that simply didn't mesh with their organization's reality — like trying to teach calculus to someone who's still mastering addition. The right model isn't necessarily the one in the latest business bestseller; it's the one that removes friction between your teams and the value they deliver to customers.

Assess Your Current Reality

Start by taking a brutally honest look at your organization as it exists today, not as you wish it existed in your strategy presentations. What's your company's DNA? Are you technical wizards who live and breathe code? Are you customer experience evangelists? Your existing strengths often point to the model that will require the least organizational whiplash to implement.

I once tried to force a customer journey model onto a deeply technical team of brilliant engineers — it was like asking cats to swim laps. Sometimes the best model is the one your organization will actually adopt rather than the one that looks prettiest in your management consulting deck.

Consider Your Product Lifecycle Stage

Early-stage products often benefit from the focus and flexibility of outcome-based teams. When launching your MVP, you'll need teams obsessively focused on adoption metrics and they won't have the luxury of specialized customer journey refinement.

Mature products with established user bases might lean toward customer journey models to enhance experiences users already value. Think of it like home renovation versus new construction; different approaches for different stages.

Evaluate Your Talent Pool

The harsh reality is that certain models demand specific skills from your product managers. Outcome-based approaches require strong analytical thinkers comfortable with experimentation. Customer journey models need empathetic storytellers who can champion user needs. Functional models thrive with technical specialists who speak the language of your developers.

Take stock of your team's strengths. If you have technically brilliant product managers who couldn't create a user persona to save their lives, a functional model might be your safest bet. I learned this lesson after watching a technically-minded PM try to facilitate a customer journey mapping workshop — it was much like watching someone try to open a bottle of champagne with a spoon.

Test Before You Commit

If possible, consider running a pilot with a subset of your organization before rolling out a new model company-wide. It's the organizational equivalent of trying on clothes before buying — much less embarrassing than showing up to the party in something that doesn't fit.

When transitioning to an outcome-based model, start with just two teams focused on our highest-priority metrics. The learnings from this "trial run" may save you from several potential disasters during the full implementation.

Remember that no model is perfect, and the most successful organizations remain flexible, adapting their approach as their products and markets evolve. The only true failure is clinging to an organizational structure that's clearly not working because the reorganization slide deck took you three weeks to perfect.

My biggest piece of advice is to not attempt a confusing hybrid of various models. Go in with conviction, a documented agreement of how things are going to work, and involve your team(s) in the decision making. If they feel they had a part in deciding their own fate, they will do their best to help make it work.

Concepts in Action: FlowSimple

FlowSimple began like any other software startup; with three brilliant engineers and a problem to solve. Their product, a workflow automation tool for creative agencies, had gained modest traction but then plateaued before it could be a challenger to one of its bigger competitors.

When Marie Hollins joined as their new VP of Product, she discovered an organization that was anything but product-centric. Engineering built features they found technically interesting, sales promised capabilities that were impossible, and marketing created campaigns totally disconnected from what the product actually did. The executive dashboard resembled the cockpit of a 747. The team was drowning in vanity metrics while starving for insights.

"We're tracking downloads, registered users, and social media mentions," Marie observed in her first executive meeting. "But are we measuring how much our customers actually use the product?"

The CEO, formerly their lead engineer, suddenly appeared uncomfortable. "We've always focused on feature development. Our customers keep asking for more."

Marie recognized the classic symptoms of what I've called "product centricity theater" — the organizational equivalent of rearranging deck chairs on the Titanic while calling it a naval innovation strategy.

Her transformation began small, with just one team dedicated to improving user activation. Rather than announcing a dramatic reorganization, she started with a pilot program — the equivalent of trying on clothes before buying.

The activation team's mission was simple: focus solely on getting new users to experience the core value of the product within their first session. Their North Star metric became "number of workflows automated in the first week"; a direct measurement of product value rather than a vanity metric.

At first, the organization's immune system kicked in. Remember that? The VP of Sales protested when the team declined to build a feature promised to a potential enterprise client. "If we lose this deal, we miss our quarterly target," he argued.

Marie faced her first test of product centricity. Rather than automatically prioritizing the sales request, she applied what I've learned is the "So What?" test. "How would this feature materially improve our customers' experience or outcomes?" The answer was clear; it wouldn't.

Instead of building the one-off feature, the team focused on making the core workflow creation experience dramatically simpler. Within two months, new user activation increased by 38%, and surprisingly, sales velocity improved as well. Prospects could now understand and experience the product's value immediately.

Thrilled by these results, Marie gradually expanded the plan. She implemented a modified functional model that played to the technical strengths of their team while also establishing a product council to maintain a cohesive user experience. Each product manager became responsible for a specific domain — analytics, workflow design, integrations — but met weekly to review how their areas impacted the overall customer journey.

The most contentious change came with the shift in measurements. The CFO initially dismissed the new product metrics as "cute distractions from real business numbers." But when Marie showed how their North Star metric predicted revenue growth with 85% accuracy two quarters in advance, even the finance team became believers.

One year into the transformation, FlowSimple looked remarkably different. The product roadmap, influenced by customer data and mapped to business outcomes, became their central planning document. Marketing synchronized campaigns with feature releases. Sales stopped selling futures and started highlighting genuine product features.

Marie's proudest moment came during a board meeting when the CEO, unprompted, presented their North Star metric before diving into the financial results. "Our product is now selling itself," he explained as subscription renewal rates hit record highs. "And when that happens, everything else becomes easier."

The transformation wasn't without casualties. Two executives who couldn't adapt to the new product-centric reality left the company. Several pet projects were killed after failing the "So What?" test. And there was a difficult quarter when revenue temporarily dipped while they rebuilt some weak foundations.

Marie saw that FlowSimple made decisions differently than it once did. Customer insights drove strategy rather than merely informing it. Teams moved with autonomy toward clear outcomes instead of blindly executing predefined outputs. And most tellingly, when a bug was discovered in the workflow editor, engineers fixed it immediately without being asked — they had watched actual customers struggle and couldn't bear to let it continue.

FlowSimple had become genuinely product-centric. And as this chapter promised, they knew it when they got there.

Now that the organization is truly product-centric, the industry established roles of product management leadership in the next chapter will more easily fit. Marie can now start forming teams that work together more effectively, she has the support of executives, and she can create amazing features by collaborating across different departments.

Execution Essentials

Embracing True Product Centricity

- Develop "customer obsession" by having all team members, including engineers and executives, regularly observe users interacting with your product.

- Apply the "So What?" test to every feature proposal: "How does this materially improve the customer experience?"

- Identify a North Star metric (like Spotify's "time spent listening") that captures the core value your product delivers.

- Allow product insights to bubble up from customer data rather than imposing top-down mandates.

Avoiding Common Pitfalls

- Watch for "product centricity theater" where teams use the buzzwords without changing how decisions are made.

- Beware the "Matrix of Confusion" where employees have conflicting priorities from multiple bosses.

- Avoid the "Budget Betrayal" by providing stable funding for product teams rather than project-based budgeting.

- Prevent the "Metrics Mirage" by treating product metrics as leading indicators for financial outcomes.

Selecting the Right Organizational Model

- Choose from Customer Journey (organized by user experience stages), Outcome-Based (focused on business metrics), or Functional (organized by domain expertise).

- Match your model to your product's lifecycle stage — early products often benefit from outcome-based approaches, mature products from customer journey models.

- Evaluate your talent pool's strengths — technical specialists may thrive in functional models, storytellers in customer journey models.

- Run small pilots before full implementation to test what works for your organization.

- Involve your team in the selection process to increase buy-in and commitment.

Navigating Transformation

- Start with small, empowered teams as proof points before rolling out organization-wide.

- Acknowledge that changing performance metrics will create discomfort as people's professional identities are tied to existing measures.

- Create a unified measurement method connecting product health to business outcomes.

- Establish a global roadmap that provides transparency across functional teams.

- Remember that becoming product-centric is itself a product challenge requiring an iterative, evidence-based approach.

Leadership Requirements

- Look for courage at the top — leaders who protect product teams from quarterly pressure when appropriate.

- Create clarity in organizational structure where authority aligns with accountability.

- Build metrics consistency with methods that allow everyone to speak the same language.

- Be prepared to kill leadership's pet projects if data shows customers don't care about them.

The Role
Chapter 2

Industry Established Leadership Roles

In the vast ecosystem of technology and most other companies, product management leadership roles often resemble a collection of instruments in an orchestra — each plays a distinct part, yet harmony depends on their collective alignment. As seasoned product leaders know, understanding the nuances between titles like Product Manager, Senior Product Manager, Director of Product, and Chief Product Officer isn't merely an exercise in corporate taxonomy.

Many organizations — even those considered industry giants — frequently misapply these titles or, worse, create confusing hybrid roles that blur responsibilities and reporting lines. Like trying to build a house using a map drawn by someone who's only heard rumors about architecture, these companies inadvertently create organizational structures that undermine the very products they hope to build.

And the chances are very high that you are currently employed by one of those companies. So, I found it fitting to engage in this topic, clarify the roles, and write this chapter.

Why Titles Matter More Than You Think

The distinction between a Product Manager and a Product Director isn't just about salary bands or corner offices. These role are different in scope, decision-making authority, and strategic influence. When properly understood and properly implemented, this leadership hierarchy creates clarity that cascades throughout the organization, allowing each product professional to operate within well-defined boundaries.

This chapter covers the core product leadership roles; not as abstract org chart entries but as essential archetypes that, when properly positioned, can transform product development from chaotic to coordinated.

The Ripple Effects of Misalignment

When organizations muddle product titles, they're essentially sending staff members into uncertain waters and without navigational charts. Consider a Senior Product Manager who believes they have authority to make roadmap decisions, while their Director assumes the same authority. Like two conductors attempting to lead the same orchestra, a noisy mess inevitably ensues.

These misunderstandings ripple through teams like fault lines beneath once stable ground. Engineers will receive conflicting priorities. Designers will create solutions for problems that may not align with business goals. Marketing prepares campaigns for features that change often or vanish entirely.

Consider a real case of a mid-sized software company that created a "Technical Product Director" role alongside their existing "VP of Product" position without clearly defining boundaries between them. What followed resembled a slow-motion organizational collision.

The Technical Product Director, Sara, believed her mandate included making final decisions on feature prioritization based on technical feasibility and architectural concerns. Meanwhile, Michael, the VP of Product, believed that all roadmap decisions required his final approval. Both were correct from their perspective. Their job descriptions contained enough ambiguity to support both of their viewpoints.

When the company's flagship product needed significant architecture changes to support enterprise clients, the ripple effects of this authority overlap spread rapidly. Engineers received contradictory directions. Sara pushed for a six-month refactoring effort before adding new features, while Michael insisted on pushing out features while making incremental architecture changes.

Design teams found themselves redoing work as priorities kept shifting. Marketing delayed major campaigns, uncertain which features would actually ship. The customer success team hesitated to make promises about future functionality. And the worst part: executive leadership received conflicting progress reports, eventually leading to a loss of confidence in the entire product organization.

What began as a seemingly inconsequential role and title confusion ultimately delayed the enterprise product launch by nine months. The whole organization felt miles off course.

Benefits of Proper Title Structure

Proper title delineation functions like a well-engineered traffic system. It creates clear lanes of responsibility that prevent collisions and gridlock. When a Principal Product Manager understands their remit encompasses technical architecture guidance across multiple products, while a Group Product Manager knows they oversee the strategic alignment of a product suite, each can focus on excellence within their domain rather than territory defense.

Well-designed title structures operate like the foundation of a skyscraper; invisible to casual observers but critical to supporting everything built upon them. Organizations that invest in properly defined, industry standard product leadership roles reap benefits across multiple dimensions.

First, decision velocity increases dramatically. When teams understand precisely who holds decision authority for different types of choices — from tactical feature specifications to strategic market positioning — they spend less time in consensus-seeking deliberations and more time executing. Like a kitchen with clearly designated stations and a respected head chef, the product development process moves with rhythm rather than chaos.

Second, career development becomes intentional rather than accidental. Junior product managers can see the skills and capabilities required for advancement, allowing them to deliberately cultivate these competencies. This creates something akin to a well-marked hiking trail with clear milestones, rather than forcing ambitious team members to bushwhack through organizational undergrowth.

Third, cross-functional collaboration improves as adjacent departments — engineering, design, marketing, sales — know precisely which product roles to engage for specific types of decisions. No more situations where an engineering leader must navigate a labyrinth of product titles to find decision authority for a technical architecture question.

Finally, executive leadership gains a clearer understanding of organizational capability. A properly structured product organization provides a reliable transmission system for translating strategic vision into market reality; allowing executives to diagnose capability gaps with precision rather than merely sensing that something isn't working properly.

Example: Trotter Analytics

Trotter began as a promising data analytics startup with an innovative AI platform. However, its product leadership structure was a textbook case of organizational confusion:

Their first major misstep was creating ambiguous hybrid roles. The company had both a "Technical Product Director" (Marco) and a "Chief Innovation Officer" (Vanessa) with overlapping responsibilities for product strategy. When the company needed to decide on API architecture changes, neither knew who had final authority.

The second problem was title inflation without clear differentiation. All five product team members carried "Senior Product Manager" titles regardless of experience or responsibilities, creating confusion about decision authority and reporting structures.

Lastly, Trotter had no established decision-making method. When engineering resources needed allocation between improving the existing platform or building a new mobile app, different product leaders gave contradictory directives, leaving engineers caught in the middle.

The results were predictable and painful:

Development velocity plummeted as teams awaited clear direction. Marketing prepared materials for features that engineering had deprioritized. Customer success made promises the product couldn't deliver. The situation reached crisis point when a major enterprise client threatened to leave due to repeatedly missed deadlines.

The Transformation

After a particularly difficult board meeting, Trotter's CEO hired a seasoned product executive, Rachel, as a consultant. She immediately identified the structural issues and proposed a reorganization based on industry-standard product leadership roles:

First, Rachel created a clear product hierarchy with distinct responsibilities:

- Two Product Managers would own specific product areas
- One Senior Product Manager would oversee the core platform
- One Group Product Manager would manage the team and handle cross-product initiatives
- Marco was repositioned as Director of Product, focusing on strategy and stakeholder management
- Vanessa moved to focus exclusively on long-term research initiatives

Second, Rachel implemented decision methods that specified exactly who had authority for different types of decisions; from tactical feature specifications to strategic market positioning.

Third, she established regular synchronization rituals including a weekly product council where all product leaders aligned on priorities and resolved conflicts before they reached engineering.

The Results

Within three months, Trotter's transformation showed measurable results:

As this chapter predicted, "decision velocity increased dramatically." Features that previously took months of deliberation were now approved in days because everyone understood exactly who made which decisions.

Career development became intentional rather than accidental. Junior team members could now see clear growth paths and the skills needed for advancement at each level.

Cross-functional collaboration improved significantly. When sales needed technical specifications for enterprise clients, they knew exactly which product role to engage rather than sending requests to multiple people.

Most importantly, executive leadership gained clarity on organizational capability. The CEO could now diagnose capability gaps with precision rather than sensing that "something isn't working properly."

Six months after the reorganization, Trotter Analytics successfully closed their largest enterprise deal to date. The client specifically cited the company's improved product communication and delivery reliability as decisive factors in their selection.

NOTE

Keep in mind that this is a leadership focused book. Therefore, the roles of Product Manager and Associate Product Manager (Junior) are not provided in the following roles, since they are not considered leadership positions.

It's best to simply make the assumption that non-leader product positions are nearly the same as the Senior Product Manager. The key difference is the amount of experience, trust in judgement, authority to make some decisions, and the ability to guide others.

NOTE

For reference convenience, each of the following roles will start on a right page.

Senior Product Manager

In the ever-evolving landscape of product development, few roles carry the unique blend of responsibility, influence, and pressure quite like the Senior Product Manager. Positioned at the crucial intersection of business strategy and technical execution, these professionals are the unsung heroes who transform abstract visions into tangible products that customers actually want to use — and occasionally even enjoy.

The Role

The Senior Product Manager (SPM) sits at a critical junction in the product management hierarchy; not quite at the executive level, but certainly elevated from the individual contributor realm. Like a seasoned chess player who can see multiple moves ahead, the SPM occupies a strategic yet hands-on position. They typically report to a Director of Product or VP of Product, depending on the organization's size and structure. Junior Product Managers and Associate Product Managers often report directly to them, creating a small but mighty product team focused on a specific product area or suite of features.

The SPM serves as both shield and sword; protecting their team from unnecessary distractions while cutting through ambiguity to drive product decisions forward. They're responsible for translating high-level company strategy into practical product roadmaps and ensuring those plans materialize into valuable customer experiences.

Unlike Product Directors who may focus primarily on strategy and leadership, SPMs must maintain close contact with the product itself, keeping one foot firmly planted in execution while the other steps toward broader strategic horizons.

The Person

The ideal Senior Product Manager resembles a Swiss Army knife — versatile, reliable, and equipped with multiple tools for various situations. They combine analytical rigor with creative thinking, technical understanding with business acumen, and leadership presence with individual contribution capabilities. They're comfortable with ambiguity, treating it not as an obstacle but as fertile ground for opportunity.

Successful SPMs tend to be naturally curious individuals who ask "why" as instinctively as they breathe. They're systems thinkers who can zoom out to see the big picture while simultaneously zooming in on critical details. Like a translator fluent in multiple languages, they can speak engineering-ese with developers in the morning, design-speak with UX teams at lunch, and business-ese with executives in the afternoon.

The best candidates for this role demonstrate both breadth and depth; wide-ranging knowledge across multiple domains coupled with deep expertise in at least one or two areas. They're perpetual learners who devour books, articles, and industry trends, constantly honing their craft while (sometimes) remaining humble enough to recognize they'll never know everything.

Collaboration

The Senior Product Manager's calendar resembles Grand Central Station; constantly alive with interactions spanning the entire organization. On a daily basis, they collaborate intimately with designers, engineers, and data analysts who form their core product squad. This tight-knit team functions as their primary collaborative unit, working in lockstep to build and refine products.

Radiating outward, they regularly engage with marketing teams to ensure appropriate positioning and messaging, sales teams to gather field intelligence and address feature requests, and customer success teams to understand user pain points and validate potential solutions.

On a weekly or monthly cadence, they interface with finance to discuss budgets and forecasts, legal for compliance considerations, and senior leadership to ensure alignment with company strategy. Like diplomats shuttling between nations, they also maintain connections with other product teams to identify integration opportunities and avoid duplication of efforts.

More occasionally, but no less importantly, they engage directly with customers through user interviews, site visits, and beta testing programs. These interactions provide the reality check that keeps product decisions grounded in genuine user needs rather than internal assumptions.

Daily Routine

The SPM's day resembles a carefully choreographed dance between tactical execution and strategic thinking:

7:30 AM   Reviews emails and Slack messages, addressing urgent issues before the workday fully begins.

8:30 AM   Engineering team daily standup — Identifying any blockers they can help remove.

9:00 AM   Leads a product prioritization session with design and engineering, evaluating new feature requests against strategic goals.

10:30 AM   Joins a customer call to understand pain points and gather feedback on recent releases.

12:00 PM   Takes a working lunch to catch up on industry news and competitor movements.

1:00 PM   Meets with data analysts to review performance metrics for recent feature launches.

2:00 PM	Collaborates with marketing on upcoming release announcements and positioning.
3:00 PM	Prepares for and delivers a roadmap presentation to stakeholders, fielding questions about timelines and priorities.
4:30 PM	Creates documentation for upcoming features, ensuring clarity for engineering and design.
5:30 PM	Sets aside time for strategic thinking; considering long-term product direction and potential innovations.
10:30 PM	Considers going to bed.

Throughout this rhythm, they remain flexible; ready to handle the unexpected questions, impromptu meetings, and occasional emergencies that punctuate product development.

My last remark in the above example and about going to bed at 10:30 PM is very real with a lot of SPMs, but doesn't have to be. I always encourage my product managers to promote a better work-life balance. If we're not getting enough done during normal work hours, then we'll talk about that – not absorb it into our personal lives. A lot of employees, coming from other organizations, are surprised when finding out that I'm very serious about this policy and compassion — and that's a shame.

Challenges

The SPM navigates challenges like a ship's captain through stormy seas. They face constant pressure from multiple directions — executives pushing for faster delivery, engineers advocating for technical debt reduction, customers demanding new features, and competitors threatening market position.

Decision fatigue represents a significant occupational hazard. With dozens of judgment calls required daily, each carrying potential consequences for users and business outcomes, mental exhaustion becomes a real concern. The SPM must develop robust decision-making methods to maintain consistency without becoming robotic.

Political savvy becomes essential when navigating competing priorities across teams. When marketing's timeline conflicts with engineering's capacity, or when sales promises features that aren't on the roadmap, the SPM must negotiate solutions that preserve relationships while protecting product integrity.

Perhaps most challenging is the accountability/authority gap. SPMs bear significant responsibility for product outcomes but often lack direct authority over all resources needed for success. They must learn to influence without authority, guiding teams toward desired outcomes through persuasion rather than mandate.

Skills

The successful SPM has cultivated a diverse skill portfolio:

- Technical understanding sufficient to have credible conversations with engineers, even if they can't code themselves. They grasp system architecture concepts, appreciate technical constraints, and recognize implementation complexity.

- Business acumen that connects product decisions to revenue impacts, market expansion opportunities, and competitive differentiation. They can build financial models to evaluate feature ROI and understand how product metrics connect to business outcomes.

- Communication skills that would make a diplomat proud — they craft compelling narratives around product vision, translate technical concepts for non-technical audiences, and deliver presentations that inspire action.

- Data literacy allowing them to derive meaningful insights from analytics, conduct effective A/B tests, and identify patterns in user behavior that inform product improvements.

- Project management capabilities for tracking complex work streams, identifying dependencies, and keeping multiple initiatives on track simultaneously.

- Design thinking that puts user needs at the center of product development, balancing desirability with feasibility and viability.

- Soft skills including empathy, active listening, conflict resolution, and team leadership prove equally critical for navigating the human aspects of product development.

Salary Range & Benefits

As of early 2025, Senior Product Managers enjoy competitive compensation reflective of their pivotal role. Base salaries typically range from $140,000 to $210,000, varying by location, industry, and company size. Those in major tech hubs like San Francisco or New York command the higher end of this range, while those in emerging tech centers might see slightly lower figures.

Experience significantly impacts compensation. SPMs with 5-7 years of relevant experience generally land in the lower half of the range, while those with 8+ years often reach the upper tiers. Industry specialization also plays a role, with enterprise software, fintech, and healthtech typically offering premium compensation.

Beyond base salary, most SPMs receive bonus potential between 10-20% of their base, equity grants worth $30,000-$100,000 annually (depending on company stage), and comprehensive benefits packages including healthcare, retirement plans, and increasingly, flexible work arrangements.

Example Career Path

Consider Jamie Chen, who currently thrives as a Senior Product Manager at a mid-size B2B SaaS company. Her journey began not in product management but as a customer support representative at a small startup. This frontline experience gave her invaluable insight into user pain points and common product questions.

After two years, Jamie transitioned to a Product Analyst role, where she learned to translate customer feedback into actionable product improvements. Recognizing her aptitude for product thinking, she pursued an MBA while working, enhancing her business acumen and strategic thinking capabilities.

Post-MBA, Jamie secured an Associate Product Manager position at a larger company, where she managed a specific feature set within a broader product. Her performance earned her a promotion to Product Manager after 18 months, expanding her responsibility to an entire product within the suite.

Three years later, Jamie leveraged her accumulated experience to land her current Senior Product Manager role, where she now leads a team of junior PMs and oversees a critical product line. Looking ahead, Jamie aspires to a Director of Product role within two years, with long-term ambitions of becoming a Chief Product Officer.

P.S.: I have no more than a Bachelor of Science in Management Information Systems. I only advocate getting an MBA for younger PMs, who are fairly new to business and product management. It could give them an edge by boosting their business acumen, if they're not coming from an engineering environment and/or a long-term product career, like me.

The same with certifications. A lot of employers seek candidates with a CSPO (Certified Scrum Product Owner) certification. What real use does that do for someone who has already owned a plethora of products and who has gotten the best on-the-job training possible? None.

10 Interview Questions

1. Tell me about a product initiative you owned from conception to launch. What metrics did you use to measure its success?

2. How do you prioritize features in your product backlog? Walk me through your method with a specific example.

3. Describe a time when you had to make a data-driven decision that went against your initial instinct. What was the outcome?

4. How do you collaborate with engineering teams to resolve technical constraints while maintaining your product vision?

5. Share an example of how you've incorporated user feedback to improve a product. What was your process for collecting and analyzing that feedback?

6. Tell me about a time when you had to pivot your product strategy based on market changes or competitive pressure.

7. How do you communicate product roadmaps to different stakeholders (executives, engineers, sales, customers)? How do you tailor your message?

8. Describe your experience working with designers to create user experiences that balance business needs with user needs.

9. Tell me about a product or feature that failed. How did you identify the failure, what did you learn, and how did you apply those lessons?

10. How do you stay informed about market trends and user behaviors in your product area? Give me an example of how this knowledge influenced a product decision.

Group Product Manager

Seated at the intersection of strategy and execution, the GPM orchestrates product teams, juggles stakeholder expectations, translates business needs to technical requirements, and somehow finds time to maintain their sanity through it all. Read on to discover the fascinating world of this pivotal product leadership role — where spreadsheets meet psychology, and corporate politics meet genuine innovation.

The Role

The Group Product Manager (GPM) occupies a critical middle management position in the product organization hierarchy, bridging the gap between individual contributor Product Managers and higher-level product leadership. GPMs typically report to a Director of Product or VP of Product, depending on the organization's size and structure. They lead and manage a team of Product Managers who, in turn, may have Associate Product Managers reporting to them.

This role is very similar to a Senior Product Manager in most ways. Where they differ is in their people management responsibility. Where SPMs advise and lead other PMs, they're not typical directly managing subordinates. GPMs typically do. They're responsible for standing up teams and creating the product management model required (see Chapter 1).

In smaller organizations, a GPM might oversee product managers across multiple product lines, while in larger organizations, they may focus on a specific product area or domain. The essence of the role involves both strategic thinking and operational execution; translating high-level company vision into actionable product strategies while ensuring their teams deliver successfully against these strategies.

The GPM role represents the first substantial management step in a product leader's career journey. Unlike individual contributor Product Managers who focus primarily on their specific product areas, GPMs must balance hands-on product work with people management responsibilities. They are accountable not just for product outcomes but for the growth and performance of the product managers they supervise.

The Person

The ideal Group Product Manager embodies a blend of seemingly contradictory traits. They are simultaneously strategic visionaries and practical executors, empathetic leaders and data-driven decision makers. This person thrives in ambiguity and possesses the rare ability to see both the forest and the trees; understanding how individual product decisions connect to broader business objectives.

Successful GPMs typically demonstrate a healthy obsession with customer problems rather than technological solutions. They possess strong emotional intelligence, enabling them to navigate complex organizational dynamics while building trust with cross-functional partners. The most effective GPMs combine analytical rigor with intuitive product sense, using data to inform decisions while recognizing when to trust their instincts.

Perhaps most importantly, the GPM must be comfortable with influence rather than authority. Despite their title, they rarely have direct control over the resources (technical and marketing teams, etc.) needed to bring products to life. Instead, they succeed through relationship building, persuasive communication, and the creation of shared vision that motivates diverse stakeholders to align their efforts.

Collaboration

The Group Product Manager operates at the nexus of multiple organizational interfaces, making collaboration a defining aspect of the role. On a daily basis, they interact closely with their direct reports (Product Managers), cross-functional teams (Engineering, Design, Data Science), and peer-level managers from other departments.

Regular collaborators include Engineering Managers who oversee the technical resources building the product, Design Managers responsible for user experience, and Marketing leaders planning go-to-market activities. GPMs also frequently engage with the Sales organization to understand customer feedback and with Customer Success teams to optimize product adoption.

On a less frequent but equally important basis, GPMs collaborate with Legal and Compliance teams on regulatory matters, Finance on budgeting and forecasting, and senior executives during strategic planning cycles. They may also interface with external partners, vendors, and occasionally key customers during strategic initiatives.

The most successful GPMs develop a reputation as connectors, creating bridges between organizational silos and facilitating communication across disparate teams; a skill that becomes increasingly valuable as they progress to more senior leadership roles.

Daily Routine

The day in the life of a Group Product Manager resembles a high-stakes juggling act performed while walking a tightrope. Here's a glimpse into their typical schedule:

8:00 AM   Quick review of key metrics dashboards and overnight customer feedback

8:30 AM   Stand-up with one of their product teams

9:00 AM   One-on-one coaching session with a Product Manager

10:00 AM   Cross-functional prioritization meeting with Engineering and Design

11:00 AM   Meeting with Data Science to review A/B test results

12:00 PM   Lunch (often working while eating at desk, despite best intentions)

1:00 PM   Product strategy session with leadership team

2:00 PM   Customer interview or feedback session

3:00 PM   Another one-on-one with a different Product Manager

4:00 PM   Roadmap review with stakeholders

5:00 PM   Catch up on email, Slack messages, and documentation

7:00 PM   Late evening review of presentations for tomorrow's meetings

What makes this schedule particularly challenging is the constant context-switching between strategic discussions and tactical problem-solving, between people management and hands-on product work. The most effective GPMs develop rituals to maintain focus amidst the chaos, carving out protected time for deep thinking about product strategy.

Most GPMs I have worked with also have executive assistants. Those valuable individuals help to balance the schedule, answer email messages on a timely basis, setup and facilitate employee reviews, handle HR paperwork, and make sure their leader is home in time for his or her anniversary.

Challenges

The Group Product Manager role comes with a unique set of challenges that test even the most seasoned product professionals. Chief among these is the tension between breadth and depth; needing to stay sufficiently informed about multiple product areas without micromanaging their teams. Many new GPMs struggle to find this balance, either remaining too hands-on or swinging too far in the opposite direction.

Political navigation represents another significant challenge. GPMs often find themselves advocating for resources in zero-sum environments, negotiating competing priorities between different stakeholders, and managing the delicate relationship between product and engineering organizations. Their position in the middle of the hierarchy means they must both implement decisions from above and absorb pressure from below.

Perhaps the most existential challenge is the constant battle against short-term thinking. While executives push for quarterly results and engineers seek technical elegance, the GPM must maintain focus on sustainable product value creation. This often means saying "no" to good ideas that don't align with the strategic direction; a necessary but frequently unpopular stance.

Finally, GPMs face the challenge of their own career evolution. The skills that made them successful individual contributors — detailed product knowledge, hands-on execution — must be partially released to make room for new competencies in people development, strategic thinking, HR skills, and organizational leadership.

Skills

A successful Group Product Manager possesses a diverse toolkit that spans both hard and soft skills. They role must:

- Orchestrate multiple products simultaneously with a cohesive roadmap that aligns with company goals while anticipating market changes. Develop strategies that consider both short-term wins and long-term product portfolio growth.

- Hire, develop, and lead exceptional product managers. Identify talent, address performance gaps, delegate effectively, and create an environment where product teams can excel independently while maintaining alignment.

- Communicate complex product concepts to diverse audiences, particularly executives. Translate technical details into compelling business narratives that demonstrate value and secure necessary resources and support.

- Build metrics methods across products with clear success criteria. Make data-driven decisions based on quantitative and qualitative evidence rather than opinions. Know which metrics truly matter for each product's stage and purpose.

- Coordinate engineering, design, marketing, sales, and customer support across multiple product lines. Manage dependencies, allocate limited resources effectively, and resolve cross-functional conflicts to maintain momentum.

- Understand market dynamics, competitive landscapes, and revenue models. Anticipate industry trends and customer needs before they become obvious. Connect product strategy directly to business outcomes and company financial objectives.

Salary Range & Benefits

As of early 2025, Group Product Managers command competitive compensation reflecting their pivotal role. Base salaries typically range from $160,000 to $220,000, with variations based on location, industry, and company size. Technology giants and well-funded startups often reach the higher end of this range, while more traditional industries may trend toward the lower end.

Total compensation packages frequently include significant additional components. Annual bonuses of 15-25% of base salary are standard, with some organizations offering profit-sharing or spot bonuses for exceptional performance. Equity grants — particularly valuable in growth-stage companies — can range from $50,000 to $150,000 annually (vesting over 4 years), though the actual value depends heavily on company performance.

Experience level naturally influences compensation. Entry-level GPMs with 5-7 years of product experience might start near the lower end of the range, while those with

10+ years of experience and proven leadership ability can command premium packages; sometimes exceeding $300,000 in total annual compensation at top-tier companies.

Beyond direct compensation, GPMs typically receive comprehensive benefits including health insurance, retirement plans, and increasingly, wellness stipends and flexible work arrangements. As companies compete for product talent, these peripheral benefits continue to expand in creative directions.

Example Career Path

Consider the journey of Priya Sharma, whose path to becoming a Group Product Manager illustrates the typical progression through product management ranks. Priya began her career as a software engineer at a mid-sized online payroll company, where she developed a reputation for asking insightful questions about user needs and business requirements.

After three years of engineering work, Priya made a lateral move to become an Associate Product Manager, taking a slight pay cut in exchange for the career pivot. Her technical background proved invaluable as she quickly demonstrated an ability to translate between business stakeholders and engineering teams.

Within 18 months, Priya was promoted to Product Manager and given responsibility for a core feature set within the company's main offering. Over the next four years, she expanded her influence, taking on increasingly complex product initiatives and informally mentoring new APMs joining the team.

When the opportunity arose for a Group Product Manager role, Priya was selected over external candidates due to her deep product knowledge and demonstrated leadership abilities. As GPM, she now oversees three Product Managers and shapes strategy for an entire product line representing approximately 30% of company revenue.

Looking ahead, Priya sees two potential paths: continuing upward in product leadership toward a Director or VP role, or potentially exploring a General Manager position with full P&L responsibility for a business unit. Her GPM experience provides valuable preparation for either direction, combining the strategic vision and execution skills required for senior leadership.

10 Interview Questions

When evaluating candidates for a Group Product Manager role, the following questions help assess their readiness for this pivotal position:

1. Can you walk me through how you've managed the product roadmap for multiple product lines simultaneously, and how you prioritized competing initiatives?

2. Tell me about a time when you had to align multiple stakeholders with conflicting priorities toward a common product vision. How did you approach this?

3. How do you measure success for your product teams? What metrics have you implemented and how did you communicate them across the organization?

4. Describe a situation where you had to make a difficult trade-off between short-term gains and long-term product strategy. What was your decision-making process?

5. How have you handled the transition of a product from concept to market? What challenges did you face and how did you overcome them?

6. Tell me about your experience mentoring and developing junior product managers. What's your approach to building high-performing product teams?

7. Can you share an example of when you identified a new market opportunity and how you validated it before committing resources?

8. How do you collaborate with engineering teams to ensure technical feasibility while still pushing for innovation? Give me a specific example.

9. Describe a product launch that didn't go as planned. What did you learn and how did you adapt your approach for future launches?

10. How do you balance customer feedback with your product vision? Can you give an example of when you chose to pursue a direction that wasn't directly requested by customers?

Director of Product Management

The Director of Product Management shepherds a team of product managers through the wilderness of stakeholder demands, market shifts, and engineering realities, all while translating executive dreams into practical roadmaps. Think of them as part diplomat, part strategist, part therapist, and occasionally, part miracle worker when deadlines approach and resources dwindle. Their caffeine intake directly correlates with the number of conflicting priorities they navigate each day.

The Role

The Director of Product Management typically sits at the middle-management level of the product organization hierarchy, nestled between the VP of Product or Chief Product Officer above and a team of Product Managers below. This pivotal position manages 4-8 Product Managers, each of whom may oversee different product lines, features, or market segments.

In smaller organizations, the Director might report directly to the CEO or COO, while in larger enterprises they usually report to a VP of Product or similar executive. The Director translates high-level company strategy into actionable product direction, empowering their team to execute while shielding them from the chaos that can rain down from the executive suite.

Unlike individual contributors who focus on specific product areas, the Director must maintain a broader perspective across the entire product portfolio while ensuring alignment with overall business objectives. They don't just build products. They build the systems, processes, and team dynamics that enable great products to emerge consistently over time.

The Person

The ideal Director of Product Management embodies a rare blend of analytical thinking and emotional intelligence. They're strategic visionaries who can also dive into the tactical details when necessary; equally comfortable discussing five-year market trends with executives and debating user interface details with designers.

These individuals typically demonstrate a healthy obsession with customer problems, balanced with business pragmatism. They possess the confidence to make difficult decisions with incomplete information, yet remain humble enough to change course when evidence suggests they should. A successful Director maintains equilibrium between optimism about possibilities and realism about constraints.

Former engineers or designers who've developed strong business acumen often excel in this role, as do business-minded professionals who've cultivated deep product sense. The common thread is an insatiable curiosity about how things work coupled with genuine empathy for both users and team members. And crucially, they've mastered the art of explaining complex concepts simply; without making the simplification simplistic.

Collaboration

The Director of Product Management is a nexus of collaboration, interfacing with virtually every department in the organization. On a daily basis, they coordinate with Engineering leadership to ensure resources align with priorities, with Design to maintain cohesive user experiences, and with their own Product Management team to guide strategy and remove obstacles.

Regular collaboration extends to Marketing for go-to-market planning, Sales for feedback and enablement, Customer Success for user insights, and Finance for budgeting and forecasting. The Director also maintains an open channel with the executive team, translating business goals into product direction.

On a less frequent but equally important basis, they engage with Legal regarding compliance and intellectual property matters, HR for team building and development, and occasionally directly with key customers or partners for strategic initiatives. In many organizations, they serve as the connective tissue that joins disparate functions into a coherent, product-focused organism.

Daily Routine

A typical day in the life of a Director of Product Management might look something like:

8:00 AM — Review key metrics dashboards and overnight communications while attempting to maintain a relationship with coffee.

8:30 AM — Stand-up with Product Management team to align on daily priorities and identify blockers.

9:00 AM — One-on-one coaching session with a Product Manager working through a particularly thorny prioritization challenge.

10:00 AM — Review product prototypes with Design lead for upcoming release.

11:00 AM — Strategic planning session with Engineering leadership to assess resource allocation against quarterly objectives.

12:00 PM — Lunch doubling as informal mentoring for a junior team member.

1:00 PM — Executive leadership meeting to present product progress and align on shifting market conditions.

2:30 PM — Deep work session reviewing and refining product roadmap based on new competitive intelligence.

3:30 PM   Customer interview to validate assumptions about an emerging market segment.

4:30 PM   Crisis management meeting addressing a production issue impacting user experience.

5:30 PM   Email triage and preparation for tomorrow's activities, including documentation of decisions made throughout the day.

Between these scheduled events, the Director fields dozens of quick questions, makes countless micro-decisions, and absorbs information from a firehose of Slack messages, emails, and impromptu conversations. Their calendar resembles a game of Tetris played on expert mode.

Like Group Product Managers, directors usually have executive assistants to help them out with schedules, communication, and business administration details.

Challenges

The Director of Product Management faces numerous challenges that test both intellectual and emotional fortitude. Chief among these is the perpetual tension between short-term demands and long-term vision. While executives push for strategic initiatives and market positioning, the day-to-day fires demand immediate attention.

Resource constraints force difficult prioritization decisions that inevitably disappoint some stakeholders. Meanwhile, the Director must protect their team from political crossfires while still ensuring they remain connected to business realities. This balancing act requires both diplomatic skill and strategic conviction.

Another significant challenge lies in measuring success. Unlike sales or engineering with their clear metrics, product management effectiveness can be maddeningly difficult to quantify in the short term. Directors must establish meaningful indicators of progress that don't distort incentives or oversimplify complex product dynamics.

Perhaps most challenging is managing the psychological weight of accountability without control. The Director bears responsibility for product outcomes while depending on multiple teams to actually deliver results; a recipe for late-night anxiety that requires tremendous emotional regulation and influence skills.

Skills

A successful Director of Product Management draws upon a diverse skill set spanning business, technology, and human dynamics. Core competencies include:

- Strategic Thinking: the ability to identify opportunities and threats in the market, connect dots across disparate information sources, and formulate coherent long-term vision.

- Communication Excellence: the mastery of tailoring messages to different audiences, from technical details for engineers to business outcomes for executives.

- Organizational Leadership: the talent for building high-performing teams, establishing effective processes, and creating cultures of accountability and creativity.

- Technical Fluency: sufficient understanding of technology to have credible conversations with engineering teams and evaluate trade-offs realistically.

- Business Acumen: strong grasp of business models, financial fundamentals, and market dynamics that impact product success.

- Data Literacy: the capability to extract meaningful insights from analytics, research, and market signals to inform decision-making.

Tool proficiency typically extends beyond standard productivity suites to include product management platforms (Jira, Productboard, Aha!), analytics tools (Amplitude, Mixpanel), prototyping software, and project management systems.

Perhaps most importantly, successful Directors develop a refined judgment that helps them navigate ambiguity and make sound decisions despite incomplete information.

Salary Range & Benefits

As of early 2025, Director of Product Management compensation reflects both the criticality and scarcity of the role. Base salaries typically range from $150,000 to $250,000, with significant variation based on company size, industry, location, and the individual's experience.

Total compensation packages frequently include substantial equity components, particularly in technology companies. Annual bonuses typically range from 15-30% of base salary, tied to company and individual performance metrics. Additional benefits often include flexible work arrangements, professional development allowances, and comprehensive health coverage.

Experience heavily influences compensation. Early-career Directors (with 7-10 years of product experience) typically fall in the lower end of the range, while veterans with more than 15 years of experience and proven track records command premium packages. The most experienced Directors at top-tier technology companies can see total compensation exceeding $400,000 annually.

Regional variations remain significant despite remote work trends, with Silicon Valley, New York, and Seattle commanding premium compensation packages compared to emerging technology hubs.

Example Career Path

Ben Jacobson is the current Director of Product Management at a fairly large management consulting firm specializing in digital transformation. Ben began his career as a business analyst at a Big Four consulting company, where he discovered his knack for translating client needs into technical requirements.

After three years of building consulting expertise across various industries, he leveraged this background to transition into a Product Strategy Consultant role at the firm, where he excelled at connecting market trends with product opportunities. His ability to bridge business challenges and technical solutions earned him a promotion to Associate Product Manager within fourteen months.

As an APM, Ben took ownership of several client-facing digital tools that dramatically improved consultant efficiency, catching the attention of executive leadership who elevated him to full Product Manager. In this role, he successfully led a complete overhaul of the firm's proprietary consulting methodology platform that increased client satisfaction scores by 42%.

Following this success, Ben advanced to Senior Product Manager, where he mentored junior team members while taking on progressively more complex product initiatives spanning multiple industry verticals. After demonstrating his leadership capabilities during a critical company expansion into healthcare consulting, he was promoted to Group Product Manager with responsibility for the firm's entire digital products portfolio and a small team of PMs.

Ben's strategic thinking and team development skills eventually led to his current Director position, where he oversees all product initiatives that support the firm's consulting practices. Looking ahead, he aims to eventually move into a Partner-track role overseeing digital innovation, though he's also exploring whether his passion might ultimately lead to Chief Digital Officer or even general management positions where he can have broader impact on how consulting services are delivered in the digital age.

10 Interview Questions

When evaluating candidates for a Director of Product Management role, the following questions help assess their readiness for the position:

1. How have you structured product teams in your previous roles to maximize impact while ensuring clear ownership and accountability?

2. Tell me about your approach to product strategy. How have you aligned it with broader company objectives while still addressing market needs?

3. Describe how you've managed the transition of a product organization through a significant company pivot or strategic shift.

4. How do you balance innovation with operational excellence in your product organization? Share specific methods or processes you've implemented.

5. Tell me about a time when you had to make a difficult resource allocation decision across multiple product lines. What was your approach?

6. How do you measure the effectiveness of your product organization? What KPIs have you implemented and how did they evolve over time?

7. Describe your experience mentoring and developing product leaders. How have you approached building a strong product culture?

8. Share an example of how you've collaborated with C-level executives to shape product vision and secure buy-in for major initiatives.

9. How have you handled situations when product teams were not meeting expectations or deadlines? What steps did you take to course-correct?

10. Tell me about a time when you had to sunset a product or feature. How did you make the decision and manage the process with customers and internal stakeholders?

Vice President of Product

This pivotal leadership role requires equal parts visionary, strategist, diplomat, and therapist. They translate executive dreams into achievable roadmaps, manage both up and down the organizational chart, and serve as the bridge between what's technically possible and what's commercially viable.

The Role

The Vice President of Product sits at a critical junction in the product management hierarchy, typically overseeing multiple product lines or a significant product portfolio. In most organizational structures, this individual manages a team of Product Directors or Senior Product Managers who each own specific products or product areas. The VP of Product generally reports to the Chief Product Officer (CPO) or, in organizations without this role, directly to the CEO or Chief Operating Officer (COO).

This position represents the transition from tactical product management to strategic product leadership. While individual contributors and even product directors focus on execution and feature delivery, the VP of Product is responsible for overall product strategy, portfolio management, and ensuring alignment with company objectives. They translate the executive vision into actionable roadmaps and are accountable for the commercial success of the entire product portfolio.

The VP of Product serves as the connective tissue between the day-to-day product work and the company's strategic direction. They must balance short-term deliverables with long-term product vision while managing resources across multiple teams and initiatives. This role is particularly crucial in scaling organizations where product complexity increases exponentially with growth.

The Person

The ideal VP of Product is a rare combination of big-picture thinker and detail-oriented executor. They possess deep product intuition honed through years of hands-on experience, yet can rise above the minutiae to see strategic opportunities. These individuals are typically characterized by intellectual curiosity paired with pragmatic decision-making skills.

Successful VPs of Product demonstrate remarkable versatility — comfortable discussing technical architecture with engineering teams in the morning and presenting business cases to the board in the afternoon. They exhibit high emotional intelligence, recognizing that product management is fundamentally about people management, whether those people are customers, stakeholders, or their own team members.

These leaders tend to be naturally collaborative but decisive when necessary. They know when to seek input and when to make the call, even with imperfect information. The best VPs of Product maintain a healthy balance between confidence in their vision and humility about what they don't know. They're often described as "product-obsessed" but in a constructive way. Their passion for creating exceptional products drives organizational momentum rather than creating bottlenecks.

Perhaps most importantly, successful VPs of Product are resilient. They understand that product development is inherently messy, that priorities will shift, and that the path from concept to market rarely follows a straight line. Like a seasoned captain navigating stormy seas, they maintain composure when chaos erupts, providing stability for their teams when uncertainty is at its peak.

Collaboration

The VP of Product is the ultimate cross-functional player, interfacing with virtually every department in the organization. On a daily basis, they collaborate closely with their Engineering, Design, and Product Marketing counterparts, forming what is often called the "product trio" or "product quad" at the leadership level. These relationships are critical for ensuring product strategy translates into technical execution, user experience, and effective go-to-market activities.

Regular collaboration also occurs with the Sales and Customer Success teams to gather market feedback and address customer concerns. The VP of Product serves as the voice of the market within the product organization, synthesizing input from customer-facing teams to inform product direction.

On a weekly or bi-weekly cadence, they typically engage with Finance to discuss budgeting, resource allocation, and product performance metrics. Collaboration with Legal and Compliance ensures products meet regulatory requirements, particularly in highly regulated industries.

The VP of Product maintains ongoing communication with executive leadership, articulating product strategy and results. Less frequent but equally important are interactions with the Board of Directors, investors, industry analysts, and key customers or partners. These external relationships help shape perception of the product in the marketplace and secure buy-in for major strategic initiatives.

Perhaps the most crucial collaborative relationship is with their own product team. Effective VPs of Product create a culture of transparent communication, mentorship, and professional development within their organization. They understand that their success is ultimately determined by the collective performance of the product managers they lead.

Daily Routine

The VP of Product's calendar resembles a game of Tetris — constantly rearranging and optimizing to fit everything in. Here's what a typical day might look like:

7:30 AM Early morning review of key metrics, emails, and Slack messages before the meeting marathon begins. This quiet time is often the only opportunity for uninterrupted strategic thinking.

8:30 AM   Stand-up with the product leadership team to align on priorities and address any immediate blockers. A quick temperature check on all active initiatives.

9:00 AM   One-on-one meetings with direct reports, rotating through the team throughout the week. These sessions focus on mentorship, progress reviews, and career development.

10:30 AM   Product review meeting with a specific product team, diving deep into designs, user research findings, or feature specifications. This is where the VP provides guidance on product direction and quality.

12:00 PM   Working lunch with the Engineering VP to discuss resource allocation, technical debt priorities, and alignment on upcoming initiatives.

1:00 PM   Executive team meeting to share product updates and coordinate cross-functional initiatives. This is where the VP advocates for product needs and navigates competing organizational priorities.

2:30 PM   Customer or prospect call to hear feedback directly from users or understand market requirements. These conversations inform product strategy and provide reality checks on current development directions.

3:30 PM   Meeting with Data and Analytics to review product performance metrics, user behavior patterns, and experiment results. Data-driven decision making in action.

4:30 PM   Product roadmap review with stakeholders from Sales, Marketing, and Customer Success to ensure alignment on upcoming releases and messaging.

5:30 PM   Final email check and preparation for the next day, including review of any urgent issues that emerged during the day.

Between these scheduled events, the VP of Product handles impromptu requests, makes countless decisions via Slack or email, and occasionally finds time to document strategic thinking or review product specs. The most effective VPs zealously protect at least a few hours each week for deep work and strategic planning — otherwise, the urgent constantly overwhelms the important.

Challenges

The VP of Product role comes with a unique set of challenges that test even the most seasoned professionals. Chief among these is managing competing priorities and expectations from multiple stakeholders. The Sales team wants new features to close deals, Engineering needs time for platform stability, executives demand innovation, and customers expect their specific problems to be solved yesterday. The VP of Product must make difficult trade-off decisions that inevitably disappoint someone.

Resource constraints present another perpetual challenge. There are always more ideas than capacity to execute them, requiring the VP to make tough calls about what doesn't get built. This scarcity extends beyond engineering resources to include design capacity, research bandwidth, and even their own attention.

The political dynamics can be particularly nuanced. As the owner of the "what" and "why" of product development, the VP of Product must influence without direct authority, especially when working with engineering and other departments. They must navigate competing agendas and build coalitions to move initiatives forward.

Maintaining strategic focus amid daily firefighting presents another significant challenge. The role naturally attracts tactical issues that can consume the calendar, making it difficult to allocate time for long-term thinking and planning.

Perhaps the most insidious challenge is the accountability-authority gap. The VP of Product is held accountable for product success but often lacks complete authority over all factors that determine outcomes. They depend on engineering delivery, marketing execution, sales effectiveness, and myriad other factors outside their direct control.

Finally, there's the internal challenge of imposter syndrome. Given the breadth of knowledge required — from technical understanding to business acumen to design thinking — even experienced VPs sometimes question if they're qualified to make the consequential decisions that land on their desk. The best ones use this self-awareness as motivation for continuous learning rather than paralysis.

Skills

- The VP of Product must excel in developing and articulating compelling product vision and strategy, with the ability to translate business goals into actionable roadmaps. Building and mentoring high-performing product teams requires exceptional leadership, as does making difficult prioritization decisions using structured methods when resources are constrained.

- Excellence in storytelling and presentation to diverse audiences enables the VP to communicate effectively across the organization. Their persuasive capabilities help build consensus among departments, while active listening skills allow them to truly understand stakeholder needs. Strong written communication is essential for documenting strategy and direction that guides the entire product organization.

- Data literacy and comfort with product metrics form the analytical foundation of the role. The VP needs sufficient understanding of software development processes and technical constraints to have credible conversations with engineering teams. Experience with product management tools like Jira, Aha, or ProductBoard facilitates workflow, while knowledge of user research methodologies ensures product decisions are grounded in customer needs.

- Financial literacy allows the VP to develop compelling business cases and manage budgets effectively. Market awareness and competitive analysis capabilities help identify opportunities and threats. Understanding of pricing strategies and revenue models directly impacts product profitability, while experience with go-to-market planning bridges the gap between product development and commercial success.

- Deep empathy for user needs and pain points ensures products solve real problems. The ability to identify market opportunities before competitors provides strategic advantage. Experience gathering and synthesizing customer feedback creates a continuous improvement loop, while the skill to define and validate product-market fit is critical for new initiatives.

- Process design and optimization expertise allows the VP to scale product operations efficiently. Resource allocation and capacity planning skills help maximize output from limited teams. Risk management strategies minimize potential disruptions, while the ability to define and track meaningful success metrics ensures accountability and continuous learning across the product portfolio.

Salary Range & Benefits

As of early 2025, the compensation package for a VP of Product reflects the critical nature of this role to organizational success. Base salaries typically range from $180,000 to $300,000, varying significantly based on company size, industry, location, and the scope of product portfolio under management.

In high-cost markets like San Francisco, New York, or London, base salaries trend toward the upper end of this range, while smaller markets or earlier stage companies might offer compensation closer to the lower end. However, base salary is just one component of the total compensation package.

Equity forms a substantial portion of VP-level compensation, especially in technology companies and startups. VPs of Product can expect equity grants valued between 0.25% and 1.5% at earlier-stage companies, or significant restricted stock units (RSUs) at public companies, typically vesting over four years.

Annual bonuses range from 20% to 40% of base salary, usually tied to both company performance and specific product metrics like revenue growth, user acquisition, or retention. Some organizations include sales commission structures for product leaders when directly tied to product revenue targets.

Benefits packages at this level typically include comprehensive health insurance, generous paid time off (averaging 4-5 weeks), retirement plans with employer matching, and increasingly, wellbeing stipends and sabbatical opportunities after tenure milestones.

Total compensation packages for experienced VPs of Product in enterprise software companies or major tech platforms can exceed $500,000 annually when combining all these elements. However, the variance is substantial. A VP of Product at an early-stage startup might accept lower cash compensation in exchange for higher equity potential, while those in more established companies tend to receive more balanced packages.

Example Career Path

Jennifer Holmes's journey to becoming VP of Product at a major energy company, illustrates a typical yet instructive career progression. Jennifer began her career as a process engineer at a traditional oil and gas firm, spending four years in field operations before recognizing her interest extended beyond technical implementation to the strategic direction of energy solutions.

She made her first career pivot by moving into a Product Specialist role within the company's emerging renewable energy division, where her engineering background provided credibility with technical teams. After three years mastering the fundamentals of product development in the energy sector, Jennifer sought broader experience by joining a fast-growing clean tech company as a Senior Product Manager.

The innovative environment accelerated her growth, forcing her to wear multiple hats and gain exposure to business strategy, regulatory compliance, and stakeholder management. When her company formed a strategic partnership with a larger energy corporation, Jennifer was promoted to Product Director, managing a team of five product managers and overseeing an entire portfolio of sustainable energy solutions.

Over the next four years, she expanded her influence by championing cross-functional initiatives and developing expertise in digital transformation for traditional energy infrastructure. Her breakthrough came when she led a new smart grid product initiative that reduced operational costs by 22% while improving reliability metrics across pilot deployments.

This success positioned her for the VP of Product role at the company, where she now manages a team of four Product Directors and eighteen Product Managers across six product categories spanning both traditional and renewable energy solutions. Jennifer credits her technical foundation, deliberate skill-building in business and leadership, and willingness to navigate complex regulatory environments with her advancement.

Looking ahead, Jennifer sees two potential paths: the CPO route, which would require deepening her experience across emerging energy technologies including hydrogen and storage solutions, or the General Manager/COO path, which would leverage her growing business acumen in the rapidly evolving energy sector. She's deliberately seeking experience in international markets and public-private partnerships to prepare for either trajectory.

Jennifer's advice to aspiring VPs reflects her own journey: "Build domain expertise early, but stay adaptable as the industry evolves. Learn to balance technical feasibility with commercial viability. Develop fluency in both regulatory methods and market dynamics. And most importantly, find opportunities to bridge traditional and innovative approaches. The energy transition demands people who can speak both languages."

10 Interview Questions

1. Describe a situation where you had to make a significant product prioritization decision that disappointed an important stakeholder. How did you approach the decision, and what was the outcome?

2. Walk me through how you've structured and evolved your product organization as the company scaled. What principles guided your organizational design decisions?

3. Tell me about a time when you needed to pivot a major product strategy. What signals indicated the need for change, how did you build consensus, and how did you manage the transition?

4. How have you approached product portfolio management? Describe your method for resource allocation across existing products versus new initiatives.

5. Share an example of how you've developed a product leader on your team. What was your approach to mentorship, and how did you measure their growth?

6. Describe your process for setting and communicating product vision. How do you ensure this vision cascades effectively throughout the organization?

7. What metrics do you consider most important for measuring product success, and how have you implemented systems to track them? Give specific examples from your experience.

8. Tell me about a significant conflict between Product and Engineering that you had to resolve. What was the nature of the disagreement, and how did you address it?

9. How do you balance data-driven decision making with intuition in product development? Share an example where you had to rely primarily on one approach over the other.

10. Describe your experience managing a product through a complete lifecycle from conception to sunsetting. What key lessons did you learn about product lifecycle management?

Chief Product Officer

The Chief Product Officer (CPO) serves as the strategic leader of product development and management within an organization. This executive role bridges business objectives with customer needs, overseeing the entire product lifecycle from conception to market success. The CPO drives innovation while managing cross-functional relationships, aligns product roadmaps with company vision, and makes critical decisions about resource allocation and product direction. This section explores the multifaceted responsibilities, required competencies, and career trajectory of this increasingly vital C-suite position.

The Role

The Chief Product Officer sits at the pinnacle of the product management hierarchy, typically reporting directly to the CEO or occasionally to the Chief Operating Officer in larger organizations. This C-suite position oversees the entire product organization, with VPs of Product, Directors of Product, and Product Managers forming the reporting chain below. The CPO serves as the ultimate arbiter of product strategy, translating company vision into executable roadmaps while balancing business objectives with customer needs.

Unlike lower-level product roles that might focus on specific features or product lines, the CPO takes a holistic view of the entire product portfolio. They're responsible for making the difficult decisions about resource allocation, determining which products deserve investment and which should be sunset. In mature organizations, the CPO may oversee not just traditional product managers but also product design, user research, and sometimes product marketing functions, creating a unified product experience from conception to market.

The Person

The ideal Chief Product Officer combines the vision of a founder with the pragmatism of an operator. They're comfortable with ambiguity yet decisive when clarity is needed. Successful CPOs tend to possess a rare blend of technical understanding, business acumen, and emotional intelligence.

These leaders typically have a track record of shipping successful products and scaling teams. While many come from product management backgrounds, others transition from engineering, design, or even marketing roles, bringing diverse perspectives to product leadership. What unites effective CPOs is their ability to inspire others with product vision while remaining grounded in market realities.

The best CPOs are renaissance professionals. They are adaptable enough to dive deep into technical discussions in the morning, persuasive enough to pitch to the board in the afternoon, and empathetic enough to mentor struggling product managers in between. They understand that product management is fundamentally about making tradeoffs and have developed the courage to make unpopular decisions when necessary.

Collaboration

The CPO's calendar resembles a United Nations assembly of stakeholders. Daily collaborations typically involve the executive team, especially the CEO, CTO, and CMO. The relationship with the CTO is particularly critical, as the line between product and technology strategy becomes increasingly blurred in modern organizations.

Regular interaction with the product leadership team, including VPs and Directors, ensures alignment on execution. Customer-facing teams like sales and customer success provide vital market feedback, while data and analytics teams help validate assumptions with concrete metrics.

On a less frequent but equally important basis, CPOs engage with the board of directors, key customers, industry analysts, and potential acquisition targets. In publicly traded companies, they may participate in earnings calls or investor meetings when major product announcements are imminent.

Daily Routine

A typical day in the life of a CPO might look something like this:

7:30 AM	Review overnight product analytics dashboards and customer feedback
8:00 AM	One-on-one with CEO on quarterly product strategy alignment
9:00 AM	Product leadership team meeting to track progress on key initiatives
10:30 AM	Review design concepts for upcoming flagship feature
11:30 AM	Lunch with potential strategic partner to discuss integration opportunities
1:00 PM	Quarterly business review with finance team on product P&L performance
2:30 PM	Crisis meeting about production issue affecting key customers
3:30 PM	Preparation for upcoming board meeting presentation
4:30 PM	One-on-ones with direct reports
6:00 PM	Evening review of competitive intelligence reports and industry news

What's notably absent from this schedule is actual time to think; a luxury most CPOs create by scheduling early mornings or late evenings for strategic contemplation away from the constant demands of the organization.

Challenges

The CPO role is fraught with unique challenges that test even the most seasoned executives. Foremost among these is managing the inherent tension between short-term business pressures and long-term product vision. When the CEO demands immediate revenue growth while the engineering organization needs stability to address technical debt, the CPO must navigate these competing priorities without alienating either side.

Political challenges abound, particularly with other C-suite executives who may view product decisions as encroaching on their territory. CTOs may resist product-driven architectural changes, while CMOs might push for features that are marketable but not valuable. The CPO must build alliances without compromising product integrity.

Another significant challenge is scaling the product organization itself. As companies grow, maintaining consistent product practices across teams becomes increasingly difficult. The CPO must institute processes that provide sufficient guidance without stifling innovation or slowing execution.

Perhaps most challenging is maintaining customer empathy as one becomes more removed from day-to-day user interactions. The most effective CPOs find ways to stay connected with real customers despite their elevated position in the organizational hierarchy.

Skills

- Strategic Vision: The ability to create and articulate a compelling product direction that aligns with company objectives while anticipating market trends and customer needs. Effective CPOs can translate abstract business goals into concrete product strategies that teams can execute against.

- Business Acumen: A deep understanding of business models, market dynamics, and financial implications of product decisions. The CPO must evaluate opportunities through both customer value and business viability lenses, making resource allocation decisions that maximize return on investment.

- Technical Fluency: Sufficient technical knowledge to evaluate feasibility, understand architectural implications, and make informed tradeoffs. While not necessarily coding themselves, CPOs must speak the language of engineering to build credibility and foster productive collaboration.

- Cross-functional Leadership: The skill to influence without authority across diverse organizational functions from engineering to marketing to sales. CPOs orchestrate alignment among teams with different priorities and perspectives, creating cohesion around product direction.

- Communication Excellence: The capacity to articulate complex product concepts to varied audiences, from engineers to investors.

CPOs translate technical details for executives while making business strategy accessible to product teams, serving as a communication bridge across the organization.

- Decision-making: Comfort with making high-stakes decisions with incomplete information under time pressure. Effective CPOs develop methods that balance data-driven analysis with experience-based intuition, taking calculated risks while managing potential downsides.

- Change Management: Experience guiding organizations through product transformations, whether introducing new methodologies, pivoting strategies, or reorganizing teams. CPOs navigate resistance to change by building stakeholder buy-in and demonstrating early wins.

- Customer Empathy: A genuine understanding of and advocacy for user needs throughout the product development process. Despite being removed from day-to-day customer interactions, successful CPOs maintain connection to the market through regular customer engagement.

- Data Literacy: The capability to derive meaningful insights from complex product metrics and market data. CPOs establish measurable objectives for product performance and create feedback loops that inform iterative improvement.

- Team Development: The talent for building and mentoring high-performing product teams. CPOs identify, attract, and retain product management talent while creating career paths that develop the next generation of product leaders.

Salary Range & Benefits

As of early 2025, Chief Product Officers command premium compensation reflecting their critical role. Base salaries typically range from $250,000 to $500,000, with significant variation based on company size, industry, and location. Technology companies in competitive markets like San Francisco or New York trend toward the higher end of this spectrum.

Total compensation packages become considerably more attractive when including equity, which can range from 0.5% to 2% of company shares in growth-stage companies. For established public companies, annual equity grants often exceed the base salary itself. Performance bonuses typically add another 25-50% of base salary, tied to product-specific metrics like user growth, revenue targets, or feature adoption.

Additional benefits often include comprehensive health coverage, generous paid time off, and substantial professional development allowances. In the current market, experienced CPOs with track records of successful product launches can command packages exceeding $1 million annually at well-funded companies.

Example Career Path

Dirk Schmidt is currently CPO at a large enterprise software organization. Dirk began his career as a software engineer, spending four years building backend systems before his natural curiosity about user needs pulled him toward product work. His technical background proved invaluable when he transitioned into an Associate Product Manager role focused on developer tools.

Over the next six years, Dirk advanced from Product Manager to Senior Product Manager, then to Group Product Manager overseeing the company's API platform. During this period, he complemented his engineering foundation with an executive MBA program, gaining crucial financial modeling and corporate strategy skills that expanded his perspective beyond feature development.

An opportunity to lead product as Director at a fast-growing competitor allowed Dirk to build his executive presence and strategic thinking. There, he successfully transformed a struggling product line into the company's highest-margin offering, catching the attention of his former employer. They recruited him back as VP of Product Strategy, where he managed multiple product lines and built a team of over thirty product managers.

After demonstrating his ability to align complex product portfolios with market opportunities, Dirk was promoted to Chief Product Officer. In this role, he orchestrated a major platform consolidation that reduced technical debt while accelerating the company's transition to cloud-based subscription models.

Looking ahead, Dirk sees several potential paths: stepping into a CEO role at a product-driven organization, joining the board of technology companies to provide product governance expertise, or potentially moving into a Chief Digital Officer position where he can apply his product thinking to broader digital transformation initiatives.

10 Interview Questions

1. How have you translated company strategy into actionable product roadmaps in your previous roles?

2. Describe how you've managed conflicts between immediate business needs and long-term product vision.

3. What methods do you use when deciding to invest in new products versus enhancing existing ones?

4. How have you measured the success of your product organization beyond revenue metrics?

5. Tell me about a time when you had to kill a product that teams were emotionally invested in.

6. How do you ensure customer needs remain central to product decisions as an organization scales?

7. What approaches have you found effective in collaborating with engineering leadership on technical architecture decisions?

8. Have you worked with C-suite executives and board members in the past and, if so, in what context were your meetings?

9. Describe your process for identifying and developing product leadership talent within your organization.

10. How have you managed shifting product priorities with minimal disruption to team morale and execution velocity?

Concepts in Action: Brovis Technologies

Brovis Technologies began as a scrappy startup founded by three engineers who developed a revolutionary cloud-based project management tool. In its early days, the entire product team consisted of just one person: Brian Wood, who wore all hats from gathering user requirements to planning releases.

As the company secured its Series A funding, it became clear they needed a more structured approach to product management. Brian was promoted to Senior Product Manager, the first formal product leadership role at Brovis.

Growing Pains

Brian embodied the quintessential Senior Product Manager. Like a Swiss Army knife — versatile, reliable, and equipped with multiple tools for various situations — he balanced analytical thinking with creative problem-solving while maintaining deep technical understanding of their platform.

His day mirrored the hectic schedule outlined in the text: mornings with engineering standups, afternoons with marketing and sales alignment meetings, and evenings catching up on industry trends. He faced the classic SPM challenge of "accountability without authority" — bearing responsibility for product outcomes while depending on engineering and marketing teams to execute.

As Brovis expanded its customer base from small businesses to mid-market companies, Brian found himself stretched too thin. The product complexity had outgrown a single person's capacity to manage.

Group Product Manager

The founders recognized this challenge and hired Tad Hayes as their first Group Product Manager. Tad took over management of the growing product team, which now included Brian and two newly hired Associate Product Managers.

Tad exemplified the GPM role, serving as both strategic visionary and practical executor. He created the product management model required for Brovis's next growth phase, defining clear processes for feature prioritization and roadmap planning.

His calendar became what I described earlier as "a high-stakes juggling act performed while walking a tightrope." Between coaching his product managers, aligning with stakeholders, and representing the product organization to leadership, Tad found himself in the classic GPM position of balancing hands-on product work with people management responsibilities.

When Brovis secured Series B funding aimed at enterprise expansion, the founders realized they needed to evolve their leadership structure once more.

Director of Product Management

The company recruited Elena Rodriguez as their first Director of Product Management. With experience from larger enterprise software companies, Elena implemented more formalized product practices while maintaining Brovis's innovative culture.

As the role defines earlier in this chapter, Elena served as "part diplomat, part strategist, part therapist, and occasionally, part miracle worker." She shielded her product teams from executive chaos while translating business strategy into actionable roadmaps.

Elena faced the challenge of the perpetual tension between short-term demands and long-term vision. The founders pushed for immediate features to secure enterprise contracts, while engineering advocated for architectural improvements. Elena navigated these competing priorities by creating a balanced roadmap that addressed both concerns.

Under Elena's leadership, Brovis's product team grew to fifteen people across consumer, enterprise, and platform teams.

Vice President of Product

As Brovis prepared for its IPO, the board insisted on bringing in more executive experience. They hired David Kim as Vice President of Product, who had previously scaled product organizations at two public companies.

David represented the transition from tactical product management to strategic product leadership. He restructured the product organization, with Elena and two other directors reporting to him, each overseeing specific product lines.

David built what the I earlier defined as "the ultimate cross-functional player," creating strong relationships with his counterparts in Engineering, Marketing, and Sales. He established a quarterly planning process that aligned product development with company financial targets, satisfying the board's desire for predictability while preserving room for innovation.

Like the typical VP, David faced the accountability-authority gap; being held responsible for product success while depending on engineering delivery, marketing execution, and sales effectiveness — all factors outside his direct control.

Chief Product Officer

Following a successful IPO and three years of growth, Brovis Technologies had evolved from a single product to a comprehensive project management platform with multiple integrated offerings. The CEO recognized the need for product leadership at the highest level and promoted David to become the company's first Chief Product Officer.

As CPO, David took a holistic view of the entire product portfolio, making difficult decisions about which products deserved investment and which should be sunset. He successfully navigated the inherent tension between short-term business pressures and long-term product vision, balancing shareholder expectations with product innovation.

David now occupied the pinnacle of the product management hierarchy, overseeing a complete product organization with VPs, Directors, Group PMs, and Senior PMs forming the leadership chain below him.

A Full Product Team

Ten years after its founding, Brovis Technologies had implemented the complete product leadership structure outlined in the chapter. Like the orchestra metaphor used in the opening pages, each product leadership role played a distinct part while creating harmony through their collective alignment.

Brian, now a Group Product Manager himself, reflected on how the properly defined product leadership roles had created clear lanes of responsibility that prevent collisions and gridlock. The well-designed title structure operated as I previously stated, "like the foundation of a skyscraper."

Through its evolution, Brovis had avoided the misalignment pitfalls warned about in the beginning of this chapter. Instead of creating confusion with hybrid roles or unclear boundaries, they had implemented industry-standard product leadership positions that allowed the company to scale efficiently while continuing to deliver innovative solutions to their customers.

Execution Essentials

General Principles

- Clarify boundaries between roles before implementing them; ambiguity leads to organizational collisions.

- Establish clear decision authority for each leadership level to prevent competing decisions.

- Create visible career paths to help product professionals understand growth opportunities.

- Introduce new leadership roles only when team complexity justifies them, not as status rewards.

- Remember that title structure is not merely about hierarchy but about distributing decision-making effectively.

Senior Product Manager Role

- Block at least 2 hours weekly for strategic thinking — protect this time religiously.

- Develop expertise in one business domain and one technical domain to increase your influence.

- Create clear documentation of decision-making criteria to maintain consistency and reduce fatigue.

- Build relationships with adjacent departments before you actually need their support.

- Maintain direct customer contact regardless of how busy your schedule becomes.

Group Product Manager Role

- Delegate product details while retaining strategic oversight — avoid micromanagement.

- Establish a regular cadence for team alignment to ensure cohesion across products.

- Create a formal process for cross-product dependencies to prevent communication gaps.

- Set clear expectations with executive assistants about schedule management priorities.

- Develop a mentorship approach that fits each PM's individual learning style.

Director of Product Management Role

- Implement a consistent method for measuring product success across teams.

- Build bridges between product and other leadership functions with regular coordination meetings.

- Create buffer space in roadmaps to accommodate inevitable surprises and shifts.

- Establish clear escalation paths for decisions that transcend individual product lines.

- Develop communication templates for different stakeholder groups to streamline messaging.

Vice President of Product Role

- Reserve 30 minutes daily for reading industry news and competitive intelligence.

- Create a quarterly review process to evaluate portfolio performance against strategic goals.

- Build personal relationships with key customers to maintain market connection.

- Develop simple visualization tools to communicate complex product strategies to executives.

- Establish clear role boundaries with Engineering VPs to prevent territorial conflicts.

Chief Product Officer Role

- Schedule regular time outside the office for strategic thinking without interruptions.

- Create a formal process for sunset decisions to ensure objective evaluation.

- Maintain a personal connection with frontline PMs through skip-level meetings.

- Build alliances with other C-level executives before major strategic pivots.

- Develop a balanced scorecard that connects product metrics to business outcomes.

The Role

Chapter 3

Specialized
Product Roles

I've spent my years in product management evolving from building basic web applications to overseeing complex applications that would have seemed like science fiction when I started my career. During this journey, I've observed how product management has splintered into numerous specialized disciplines, each requiring unique skills and perspectives.

When I started, we were simply "product managers" – generalists expected to handle everything from user research to roadmapping to feature specifications. Today, the field has matured and specialized, much like how medicine evolved from general practitioners to specialists with deep expertise in specific domains.

In this chapter, I'll explore six specialized product management roles that have emerged as technology has grown more complex. As a leader, you need to know about emerging job architectures in the product space. You'll probably be hiring quite of few of these characters in your near future.

Looking Ahead

As technology continues to evolve, new specialized product management roles will inevitably emerge. We're already seeing nascent specialties around augmented reality, blockchain, and quantum computing.

I believe the most successful product leaders will be those who can move fluidly between specialties, bringing cross-disciplinary insights while respecting the unique challenges of each domain. The technical PM who understands growth principles, or the AI PM with platform thinking, will have unique perspectives that drive innovation.

What remains constant across all these specializations is the core purpose of product management: creating value by identifying user needs and translating them into solutions that balance business goals with technical feasibility. The methods may vary, but this fundamental mission unites all product managers, regardless of specialty.

Whether you choose to specialize or remain a generalist, continuous learning across these domains will make you a more effective and adaptable product leader in our rapidly evolving industry.

Also keep an eye out for new product roles and titles. The addition of the AI/ML Product Manager role was recent and artificial intelligence will be ushering in a lot of new ways of building customer solutions. Surely, with all those new methods will come more specialities. We just don't know what they are yet.

NOTE

Like the previous chapter and for reference convenience, each of the following roles will start on a right page.

Technical Product Manager (TPM)

I still remember the first time I was called a "technical product manager" – it seemed redundant. Aren't all product managers supposed to understand technology? As it turns out, there's a significant difference between understanding technology conceptually and having the depth of technical knowledge required to lead highly complex technical products.

What Sets TPMs Apart

Technical product managers typically have stronger technical backgrounds than their generalist counterparts. Many come from engineering roles or hold computer science degrees. This deeper technical understanding allows them to gain credibility with engineering teams and make more informed decisions about technical trade-offs.

The difference between a general PM and a TPM became crystal clear to me early in my career. An engineering team proposed moving from a monolithic architecture to microservices. Our generalist PMs understood the business benefits – improved scalability, faster deployments – but couldn't evaluate the technical implications. Being a former software architect, I identified several critical services that would actually perform worse as microservices due to tight coupling and data dependencies. This insight saved months of refactoring work and potential performance degradation.

The TPM's Toolkit

Technical PMs speak both business and technical languages fluently. They can:

- Understand and contribute to technical architecture discussions
- Evaluate technical debt and make informed prioritization decisions
- Translate complex technical concepts for non-technical stakeholders
- Define non-functional requirements (performance, security, scalability) with precision
- Dive into metrics like CPU utilization, memory consumption, and response time

I once had a TPM who spotted a critical flaw in our performance testing methodology. While our general metrics looked good, he recognized that our API response times were hiding severe outliers that would affect our largest customers. His technical background allowed him to not just identify the issue, but work directly with engineers on determining the root cause – an inefficient database query pattern that only emerged at scale.

Senior Technical Product Manager (STPM)

As organizations mature, they often create Senior Technical Product Manager roles. These individuals combine deep technical expertise with strategic product thinking and leadership skills. Yes, I once served in this role, as well.

STPMs typically focus on:

- Defining technical strategy across multiple products or features
- Making architectural decisions with long-term implications
- Leading cross-functional collaboration on complex technical initiatives
- Mentoring junior TPMs and helping general PMs understand technical considerations
- Evaluating build vs. buy decisions for technical components

I once knew an STPM who led a transition to a zero-trust security architecture. This wasn't merely a technical decision but a fundamental shift that affected everything from user experience to our compliance posture. This guy created a three-year technical roadmap that balanced immediate security improvements with long-term architectural evolution, all while maintaining business continuity. This kind of initiative requires not just technical depth but also strategic thinking and cross-functional leadership – the hallmarks of a strong STPM.

When You Need a TPM

Not every product requires a dedicated technical product manager. But certain situations call for technical depth:

- Products with significant infrastructure or platform components
- Developer-facing APIs or tools
- High-scale systems with complex performance requirements
- Products requiring deep integration with operating systems or hardware
- Systems with strict security or compliance requirements

I've found that TPMs thrive in these environments, where technical complexity directly impacts product success. In contrast, consumer applications focused primarily on user experience may benefit more from product managers with design backgrounds.

Growth Product Manager

In 2014, I consulted with a startup that had a dedicated "growth team" – an alien concept to me at the time. "Isn't growth everyone's responsibility?" I wondered. I quickly learned that growth product management is a specialized discipline with its own methodologies, metrics, and mindset.

The Growth PM's Focus

Growth product managers optimize specific parts of the user journey to improve key metrics like acquisition, activation, retention, revenue, and referral (often called the "AARRR" or "pirate metrics" method). Unlike general product managers who might be building new capabilities, growth PMs relentlessly optimize existing user flows.

The difference became clear when I worked alongside a brilliant growth PM named Sarah. While I focused on building our core product capabilities, Sarah obsessed over our signup flow. She implemented a series of small, seemingly trivial changes – replacing dropdown menus with radio buttons, simplifying form fields, adjusting button copy – that collectively increased signup completion rates by 28%. Her changes generated more business impact in three months than some feature teams delivered in a year.

The Growth PM's Toolkit

Growth product managers use a distinct toolkit:

- A/B testing methods and experimental design
- Funnel analysis and user behavior analytics
- Rapid ideation and hypothesis generation
- Statistical analysis to evaluate experiment results
- Marketing technology integration

I once watched a growth PM transform his company's premium subscription conversion rate through methodical experimentation. Rather than relying on gut instinct about pricing or packaging, he designed a systematic series of tests exploring different price points, feature bundles, trial periods, and messaging. Each experiment built on insights from the previous one, ultimately increasing conversion by 46% while maintaining strong retention metrics.

Growth PMs think scientifically. They form hypotheses, design experiments to test them, and let data guide their decisions. They're comfortable with failure because they know that many experiments won't succeed – the key is learning quickly and iterating.

When You Need a Growth PM

Growth product managers are particularly valuable when:

- Your product has achieved product-market fit and you're focusing on scale

- There's a clear monetization model to optimize

- Your product has significant traffic for statistically valid experiments

- You're facing increased competition and need to improve conversion metrics

- User acquisition costs are rising, increasing pressure on funnel optimization

I remember interviewing an exceptional growth PM candidate who said something profound: "Feature PMs build things people want. Growth PMs ensure people get value from what's built." This complementary relationship explains why many successful companies have both traditional and growth-focused product teams.

Platform Product Manager

The term "platform" has become somewhat overused in product circles – it seems every product aspires to be a platform these days. But true platform product management is a specialized discipline focusing on creating foundational technologies that other products or developers can build upon.

The Platform PM's Focus

Platform product managers focus on creating reusable capabilities, APIs, and services that enable other teams (internal or external) to build their own products more efficiently. They think in terms of extensibility, scalability, and developer experience rather than end-user features.

I saw the power of platform thinking at a media company where they initially built separate content management systems for each publication. As they acquired more properties, this approach became unsustainable. The company's platform PM led the creation of a unified content platform that could support multiple publications with different workflows and presentation layers. This not only reduced engineering overhead but accelerated their ability to launch new publications from months to weeks.

The Platform PM's Toolkit

Platform product managers need specialized skills:

- Systems thinking and architecture design
- API design and developer experience optimization
- Creating comprehensive documentation and examples
- Balancing flexibility against complexity in extensibility mechanisms
- Versioning strategies and backward compatibility planning

The best platform PMs I've worked with share a common trait: they think in abstractions. Where a feature PM might see specific use cases, a platform PM identifies patterns that can be generalized into reusable components.

I recall a platform PM who transformed how we thought about user permissions across our enterprise product suite. Rather than building permission models for each product, she designed a unified permissions method that could express complex access patterns while remaining simple for basic use cases. This platform approach saved hundreds of engineering hours while creating a consistent experience across products.

When You Need a Platform PM

Platform product managers become essential when:

- Your organization has multiple products that share common functionality
- You're creating developer-facing APIs or SDKs

- Internal teams are duplicating efforts across products
- You're building an ecosystem where third parties will extend your core offering
- Technical complexity requires abstraction to enable efficient product development

Platform product management requires patience and long-term thinking. While feature PMs might ship visible improvements weekly, platform initiatives often take months to yield benefits. The payoff comes in the accelerated development velocity they eventually enable across the organization.

AI/ML Product Manager

When machine learning moved from research labs to production systems, it created a need for specialized product managers who understand both the capabilities and limitations of AI (Artificial Intelligence) technologies.

In today's digital landscape, artificial intelligence is evolving at breakneck speed, transforming from science fiction into business reality faster than many executives can update their LinkedIn profiles.

Consider how quickly language models have evolved — like comparing a paper airplane to a supersonic jet in just a few years. What was once impressive now seems quaint, as if comparing a calculator to a quantum computer. This rapid progression means businesses can't afford the luxury of wait-and-see approaches.

Smart product leaders are aggressively trying to understand how AI fits into their product landscape and they're rapidly expanding their teams to include these types of product managers. We've truly been given no time to wait.

The AI Product Manager's Focus

AI product managers bridge the gap between data scientists, engineers, and business needs. They identify appropriate applications for machine learning, define success metrics for models, and manage the unique development lifecycle of AI features.

The AI PM's Toolkit

AI product managers need specialized knowledge and tools:

- Understanding of fundamental machine learning concepts and limitations
- Familiarity with model evaluation metrics and validation techniques
- Ability to design data collection strategies to improve models
- Expertise in managing model performance in production
- Skills in explaining probabilistic systems to stakeholders accustomed to deterministic software

The best AI PMs excel at managing uncertainty. Traditional software is largely deterministic – given the same inputs, you get the same outputs. Machine learning systems are probabilistic by nature, which requires a different product management approach.

When You Need an AI Product Manager

AI product managers become crucial when:

- Your product relies heavily on machine learning components
- You're exploring potential applications of AI in your domain

- You need to balance model performance against other product considerations

- You're managing the ethical implications of AI systems

- You need to translate between technical AI concepts and business stakeholders

The AI/ML product manager role continues to evolve as the technology matures. The most successful ones combine technical understanding with a strong user-centered focus, ensuring that AI serves genuine user needs rather than becoming technology for its own sake.

When I write the third revision of this book, I'm sure everything I just wrote will be obsolete.

Data Product Manager

While there's some overlap with AI/ML product management, data product management focuses specifically on products that help users capture, analyze, visualize, and derive insights from data. It, too, is a relatively new role.

The Data Product Manager's Focus

Data product managers create tools that help users make sense of information. This ranges from business intelligence platforms to data pipelines, analytics tools, and visualization systems.

Data products require special attention to information architecture and cognitive load. When a user looks at a dashboard with dozens of metrics, their ability to derive meaningful insights depends entirely on how effectively you've organized and presented that information.

The Data PM's Toolkit

Data product managers employ specialized approaches:

- Data modeling and organization principles
- Information visualization best practices
- Understanding of data quality and governance
- Knowledge of analytical workflows and methodologies
- Balancing flexibility versus simplicity in analysis tools

A skilled data PM I know completely reimagined her company's executive dashboard by applying visual design principles to data presentation. She reduced the number of displayed metrics, introduced progressive disclosure for detailed information, and employed consistent visual patterns for related metrics. These changes transformed the dashboard from a confusing array of numbers into a strategic decision-making tool.

When You Need a Data Product Manager

Data product managers are essential when:

- Your primary value proposition involves helping users understand complex information
- You're building analytics tools or business intelligence platforms
- Your product generates large volumes of data that require interpretation
- Users need to make decisions based on data analysis
- You're creating data pipelines or data transformation tools

The best data PMs combine analytical thinking with strong user empathy. They understand both the technical aspects of data management and the cognitive aspects of how humans interpret information.

Concepts in Action: 30Fusion

30Fusion began as a small startup developing basic web applications for business analytics. In its early stages, the company employed just three product managers; all generalists expected to handle everything from user research to roadmapping to feature specifications.

Like the evolution described in the book chapter, 30Fusion's founders had a "we're simply 'product managers'" approach where everyone was expected to be a jack-of-all-trades. This worked well enough when their product suite was simple, but as they expanded their offerings and technology grew more complex, challenges emerged.

Growing Pains and the Need for Specialization

After securing their Series B funding, 30Fusion experienced rapid growth both in staff and product complexity. Their flagship business analytics platform was serving more customers and handling increasingly complex data processing needs. The generalist approach started showing cracks.

"We were spreading ourselves too thin," recalled Barb Sundis, VP of Product. "Our PMs were constantly context-switching between deep technical discussions and marketing strategy meetings. Something had to give."

The breaking point came when 30Fusion attempted to launch a new machine learning feature that would provide predictive analytics. Despite months of work, the feature failed to deliver value to customers. Post-mortem analysis revealed a fundamental disconnect between the product management and data science teams.

Specialized Roles Emerge

Recognizing the need for specialized expertise, 30Fusion restructured their product team to include distinct specialized roles:

Technical Product Managers (TPMs)

30Fusion hired two TPMs with engineering backgrounds to oversee their infrastructure and API products. One notable success came when senior TPM Rajiv Shah identified critical flaws in their planned microservices architecture that would have created performance bottlenecks for their largest enterprise customers.

"Rajiv spoke both business and technical languages fluently," noted the CTO. "He saved us months of refactoring work by identifying architectural issues before implementation, something our generalist PMs simply couldn't have spotted."

Growth Product Manager

As 30Fusion achieved product-market fit with their core analytics platform, they brought on Elena Gomez as their first dedicated Growth PM. Using her expertise in A/B testing and funnel optimization, Elena focused exclusively on optimizing the onboarding process and premium tier conversion rates.

Within six months, Elena's systematic experiments improved trial-to-paid conversion by 32% without changing the core product functionality. Her work generating detailed funnel analyses and implementing small, iterative improvements to the signup flow created more revenue impact than several major feature releases.

Platform Product Manager

As 30Fusion's product suite expanded to include multiple specialized analytics tools, they recognized the inefficiency of each team building similar components. They hired Jackson Williams as Platform PM to create a unified foundation that all product teams could build upon.

Jackson developed a comprehensive component library and data processing method that reduced development time for new analytics features by 60%. His platform approach meant that improvements to core capabilities automatically benefited all of 30Fusion's products, creating exponential returns on investment.

AI/ML Product Manager

When 30Fusion decided to incorporate machine learning more deeply into their analytics platform, they brought on Dr. Sophia Lin, who had both data science credentials and product experience. Unlike their previous failed ML initiative, Sophia successfully bridged the gap between data scientists and business needs.

"Sophia helped us understand that machine learning isn't magic; it's a tool with specific strengths and limitations," said the CEO. "She kept our AI features grounded in actual user problems rather than chasing the latest technical novelty."

Sophia's expertise in managing model performance and explaining probabilistic systems to stakeholders accustomed to deterministic software proved crucial for 30Fusion's successful AI transformation.

Data Product Manager

As data visualization became central to 30Fusion's value proposition, they hired Marcus Johnson as their Data Product Manager. Marcus reimagined how information was presented across their analytics dashboards by applying information visualization best practices.

His work reduced cognitive load on users by introducing progressive disclosure for detailed information and employing consistent visual patterns for related metrics. Customer satisfaction scores rose dramatically as users reported being able to derive insights more quickly from their data.

Integration Challenges

While specialization brought needed expertise, it also created new challenges of coordination and alignment. 30Fusion initially struggled with siloed teams and competing priorities among specialized PMs.

The breakthrough came when they implemented a "T-shaped" skill development program, encouraging product managers to maintain deep expertise in their specialization while developing broader understanding across adjacent domains.

Cross-functional "product pods" brought together specialists from different domains to collaborate on integrated features. For example, when developing advanced anomaly detection capabilities, they assembled a team with their AI/ML PM, Data PM, and Growth PM to ensure the feature was technically sound, visually effective, and properly introduced to users.

Results: Balanced Specialization

Three years into their specialization journey, 30Fusion had found a productive balance. They maintained specialized roles while ensuring coordination through shared methodologies and regular knowledge sharing.

The CEO reflected: "What made the difference wasn't just hiring specialists; it was creating a culture where our Technical PMs understood growth principles, our AI PMs thought about platform scalability, and everyone remained connected to our core mission of helping customers make better decisions through data."

This approach embodied my earlier point that "the most successful product leaders will be those who can move fluidly between specialties, bringing cross-disciplinary insights while respecting the unique challenges of each domain."

By embracing specialized product roles while maintaining cross-functional collaboration, 30Fusion transformed from a struggling startup to an industry leader in business analytics, demonstrating that with the right balance, specialization doesn't have to mean fragmentation.

Execution Essentials

General Product Management Evolution

- Product management has evolved from generalists to specialists as technology has grown more complex, similar to how medicine evolved from general practitioners to specialists.

- The most successful product leaders move fluidly between specialties, bringing cross-disciplinary insights while respecting each domain's unique challenges.

- What remains constant across all specializations is the core purpose: creating value by identifying user needs and translating them into feasible solutions.

- Leaders should keep an eye out for emerging product roles and titles, especially as AI introduces new ways of building customer solutions.

- Continuous learning across different product domains makes you a more effective and adaptable product leader in our rapidly evolving industry.

Technical Product Manager (TPM)

- Technical PMs should be able to identify when certain technologies (like microservices) might actually perform worse for specific use cases, saving months of potential refactoring work.

- Develop the ability to translate complex technical concepts for non-technical stakeholders to bridge the communication gap.

- Look beyond surface-level metrics to identify potential issues, such as API response times that might hide severe outliers affecting large customers.

- TPMs thrive in environments where technical complexity directly impacts product success, such as infrastructure components or developer-facing APIs.

- Senior Technical PMs should focus on defining technical strategy across multiple products and making architectural decisions with long-term implications.

Growth Product Manager

- Focus on optimizing specific parts of the user journey using the "AARRR" method: acquisition, activation, retention, revenue, and referral.

- Implement small, seemingly trivial changes to user flows that can collectively create significant business impact through continuous optimization.

- Use A/B testing methods and experimental design to validate hypotheses rather than relying on gut instinct about features or pricing.

- Think scientifically — form hypotheses, design experiments, and let data guide decisions, knowing that learning from failures is part of the process.

- Growth PMs become particularly valuable after achieving product-market fit, when focusing on scale and optimizing monetization models.

Platform Product Manager

- Think in abstractions and identify patterns that can be generalized into reusable components rather than building separate solutions for each use case.

- Focus on creating unified methods (like permissions systems) that can save hundreds of engineering hours while creating consistent experiences across products.

- Balance flexibility against complexity when designing extensibility mechanisms for your platform.

- Platform initiatives often take months to yield benefits, so practice patience and long-term thinking to see the eventual payoff in accelerated development velocity.

- Employ systems thinking to create reusable capabilities that enable other teams to build their own products more efficiently.

AI/ML Product Manager

- Bridge the gap between data scientists, engineers, and business needs by identifying appropriate applications for machine learning.

- Become skilled at explaining probabilistic systems to stakeholders who are accustomed to deterministic software.

- Excel at managing uncertainty, recognizing that AI systems are probabilistic by nature and require a different product management approach than traditional software.

- Ensure AI serves genuine user needs rather than becoming technology for its own sake by combining technical understanding with strong user-centered focus.

- Stay adaptable as the AI field evolves rapidly — what seems cutting-edge today may soon become obsolete.

Data Product Manager

- Pay special attention to information architecture and cognitive load, recognizing that user insights depend on how effectively you organize and present information.

- Apply visual design principles to data presentation, such as reducing displayed metrics and introducing progressive disclosure for detailed information.

- Balance flexibility versus simplicity in analysis tools to prevent overwhelming users with options.

- Combine analytical thinking with strong user empathy to address both technical aspects of data management and cognitive aspects of information interpretation.

- Focus on transforming confusing arrays of numbers into strategic decision-making tools through thoughtful organization and visualization.

The Role

Chapter 4

Managing
Great Teams

Surely you've heard the famous saying that people don't quit jobs, they quit managers. As a product leader for decades, I've seen this play out repeatedly. The difference between a thriving product team that ships incredible work and one that sinks into politics and frustration often comes down to one thing: leadership.

Leadership isn't about having all the answers. It's about creating an environment where your team can find them together. Think of yourself as a music teacher proudly looking upon your class of future musicians. Every student has an instrument to play, but it's you who ensures everyone plays their part in harmony.

Lane Keeping & Team Hierarchy

Every effective product team has clear lanes. Like a well-designed highway system, when everyone knows their route, traffic flows smoothly. When lanes blur, dramatic multi-car pileups happen.

As a product leader, one of my first jobs is ensuring everyone understands their specific role while appreciating how it connects to others. Product managers focus on the "what" and "why," designers on the "how it works and feels," engineers on the "how it's built," and researchers on "what users need." When these boundaries blur without intention, confusion sets in.

Consider Miguel, a product manager I once worked with. Talented but overeager, he would often jump into designing interfaces during meetings, effectively stepping on the toes of our design team. The designers, feeling undermined, gradually stopped bringing their best ideas forward. Once we clarified that Miguel should focus on articulating problems clearly and setting success metrics, leaving solution exploration to designers, team dynamics improved dramatically. Miguel could still contribute ideas, but within a collaborative method rather than prescriptively.

This doesn't mean rigid silos. The best teams have permeable boundaries; product managers who understand technical constraints, designers who grasp business goals, and engineers who empathize with user needs. The key is mutual respect and knowing when to lead versus when to support.

Staying within the product domain, the product management ladder typically includes junior product managers, product managers, senior product managers, group product managers, directors, and executives. Each step brings increased scope, autonomy, and responsibility.

What many product organizations get wrong is treating this as purely a reporting structure rather than a decision-making method. I've found the most effective teams operate with clear delegation principles, where higher positions support rather than override lower ones within their domains of ownership.

Let's start getting into some lane keeping details and techniques…

Mentorship Without Micromanagement

The relationship between Senior Product Managers and Junior Product Managers is particularly delicate. Seniors typically have enough experience to spot potential pitfalls immediately, creating a temptation to dictate solutions rather than guide discovery.

I coach my Senior PMs to operate as mentors rather than managers, even when they have formal reporting authority (which is not common). This means:

- Asking questions before offering solutions: "What factors did you consider in this decision?" rather than "Here's what you should do."

- Creating learning moments from mistakes: When a Junior PM proposed a feature without considering technical debt implications, the Senior PM walked through the architecture together rather than rejecting the proposal outright.

- Providing context, not commands: Explaining the "why" behind product principles rather than enforcing them arbitrarily.

- Giving increasing autonomy as skills develop: Starting with close guidance on smaller initiatives and gradually expanding scope as confidence grows.

I remember when a new Junior PM (APM) joined one of my teams under one of our strongest Senior PMs. The SPM resisted the urge to simply tell the APM what to build.

Instead, he involved her in user research, helped her develop her own insights, and coached her through crafting a proposal. When she presented to stakeholders, she owned that solution completely — and when it succeeded, she received full credit. The SPM's approach built her capabilities rather than dependence.

Supporting Without Undermining

Directors of Product have typically evolved beyond day-to-day product decisions to focus on strategy, team development, and cross-functional leadership. The most common mistake at this level is continuing to operate as a "super Senior PM" rather than creating space for Senior PMs to grow.

When Directors override Senior PM decisions, even with the best intentions, they create several problems:

- Undermining the Senior PM's authority with their team and stakeholders
- Creating bottlenecks where decisions require Director approval
- Stunting the Senior PM's growth in decision-making confidence
- Teaching the organization to escalate around the Senior PM

I coach Directors to adopt a "disagree and commit" mindset. If a Senior PM makes a decision the Director wouldn't have made, but it's not catastrophic, support it anyway. The learning opportunity for the Senior PM and the preservation of their authority usually outweigh the marginal benefit of an "optimal" decision.

When Lisa was promoted to Director, she struggled initially with letting her Senior PMs make calls she disagreed with. After getting some coaching on a method: unless a decision violated company values, created unacceptable risk, or fundamentally misaligned with strategy, she would support it publicly even if she'd have chosen differently. Within six months, her Senior PMs were making better decisions than she would have made herself because they combined her strategic guidance with their deeper domain knowledge.

Directors, stop micromanaging your Senior PMs. If you're still making all the calls, you're not leading — you're stifling growth. Adopt a "disagree and commit" mindset: if it's not a game-changer, let the Senior PM make the decision. Supporting them — even when you think you know better — helps them grow, and in time, they'll make decisions you wish you'd thought of first.

Coordinating Without Controlling

Group Product Managers (or similar titles like Lead PM) occupy a unique position; often responsible for a product area's overall success while supervising Senior and Junior PMs who own specific components.

The distinctive challenge at this level is balancing product area cohesion with individual ownership. I've seen Group PMs fail in two directions: becoming pure people managers disconnected from product details, or conversely, treating their teams as implementation arms for their own product vision.

Effective Group PMs:

- Own the "why" and "what" at the product area level while enabling Senior PMs to determine the "how" within their domains
- Create alignment through shared principles and objectives rather than prescriptive solutions
- Focus on removing obstacles and creating connections between product lines
- Trust Senior PMs to manage Junior PMs' day-to-day work while providing mentorship methods

Carlos, a Group PM leading a platform team, demonstrated this balance masterfully. He established clear platform principles and objectives with his three Senior PMs, then stepped back to let them determine implementation approaches with their respective teams. When conflicts arose between teams, he facilitated resolution rather than imposing decisions. His Senior PMs felt genuine ownership while still working toward cohesive platform goals.

Example: Deposito

At Deposito, a small fintech startup in Austin, Jess took over as Group Product Manager for their payments platform team. She inherited three talented Senior PMs — Dev, Maria, and Tyler — each responsible for different components of the platform.

During her first month, Jess noticed that the previous Group PM had micromanaged everything, requiring all decisions to pass through them. The Senior PMs had essentially become implementation arms for the former Group PM's vision, which killed their motivation and slowed innovation.

Jess decided to take a different approach. First, she worked with her team to establish clear platform principles and objectives that created alignment across the payment components. Then, critically, she stepped back to let the Senior PMs determine their own implementation approaches.

When Dev proposed a controversial authentication method for the paper check image deposit flow, Jess's instinct was to redirect him toward what she considered a safer option. Instead, she asked thoughtful questions about his reasoning and connected him with relevant stakeholders to refine his approach. The feature launched successfully, and Dev's ownership of the decision made him more invested in its performance.

A month later, Maria and Tyler got into a heated disagreement about API structure changes that would impact both their areas. Rather than jumping in with a solution, Jess facilitated a workshop where both PMs explained their constraints and objectives. She didn't dictate the answer but created space for them to collaborate. The solution they developed together was stronger than either of their initial proposals.

By owning the "why" and "what" at the platform level while enabling her Senior PMs to determine the "how," Jess transformed Deposito's product culture. Six months in, feature delivery had accelerated by 40%, and all three Senior PMs reported higher job satisfaction. Most importantly, when a critical competitive threat emerged, the team responded with innovative solutions that Jess herself wouldn't have conceived.

Escalating When Necessary

While respecting decision ownership is crucial, there are legitimate cases where higher-level intervention is appropriate:

- When decisions affect multiple product areas beyond the decision-maker's purview
- When significant company resources or strategic direction are at stake
- When a PM (at any level) is clearly struggling and needs support
- When team conflict requires objective mediation

The key is making these interventions transparent, supportive, and focused on long-term team health rather than expedient solutions.

When an e-commerce team proposed a change that would impact the mobile app team's user experience, tensions escalated between the Senior PMs. Rather than dictating a solution as Director, I facilitated a workshop where both teams articulated their constraints and objectives. The resulting collaborative solution was stronger than either team's original proposal, and both Senior PMs maintained ownership within their domains.

A Culture of Respectful Challenge

Hierarchy shouldn't mean blind deference. Healthy product organizations encourage respectful challenge in all directions. Junior PMs should feel comfortable questioning Senior PMs' assumptions and Directors should welcome pushback from their teams.

The difference is in how these challenges occur. Questions asked privately before public meetings, challenges framed as curiosity rather than criticism, and disagreements focused on outcomes rather than egos all preserve the decision-making method while improving the decisions themselves.

I encourage "pre-meetings" before key decisions, where team members can raise concerns in a low-stakes environment. This prevents public undermining while ensuring all perspectives are considered.

A culture that encourages respectful challenges is essential for fostering innovation, improving decision-making, and driving continuous improvement in product management.

Shared Accountability with Clear Documentation

One practical tool for navigating the product hierarchy is clear documentation of decision rights. For each product area and initiative, you should explicitly document:

- Who has decision authority (makes the final call)
- Who must be consulted (provides input that must be considered)
- Who should be informed (kept updated but not necessarily consulted)

This RACI-style approach (Responsible, Accountable, Consulted, Informed) prevents confusion and reduces political maneuvering. When everyone understands their role in decisions upfront, they can engage appropriately without feeling excluded or overridden.

Evolution of Authority

A healthy product organization isn't static. Junior PMs grow into Senior roles and Senior PMs develop toward Group or Director positions. This evolution should include intentional transfer of decision authority, not just title changes:

- Observation: Newer PMs shadow decision processes
- Participation: They contribute to decisions with guidance
- Supervised autonomy: They make decisions with review
- Full ownership: They make decisions independently with support available as needed

This progression applies between any adjacent levels in the hierarchy. The goal is always to push decision-making to the lowest appropriate level while providing support structures for growth.

Correction Without Undermining

Even in well-structured teams, mistakes happen. How leaders address these moments defines the health of the hierarchy.

If a Junior PM on your team launches a feature that causes customer confusion, the Senior PM must resist the urge to take over completely. Instead, they should work with the Junior PM to analyze what happened, develop a correction plan, and implement it together. The Junior PM will learn invaluable lessons while maintaining ownership, and the Senior PM demonstrates supportive leadership rather than punitive intervention.

The same principle applies up the chain. Directors support Senior PMs through mistakes; providing guidance without seizure of control.

Building Your Team for Harmony

Creating all of this balance starts with hiring and promotion decisions. Beyond technical skills, I look for:

- Ego awareness: Can this person support others' success without needing personal credit?

- Teaching orientation: Do they enjoy developing others or just doing the work themselves?

- Decision confidence: Can they make calls with appropriate consultation but without excessive consensus-seeking?

- Growth mindset: Do they see feedback as development rather than criticism?

These traits predict success in navigating the complex dance of team hierarchy far better than pure product or technical expertise.

Remember that the ultimate goal of your team hierarchy isn't control. It's creating a system where decisions are made at the right level, by people with the right context, supported by leaders who balance guidance with trust. When this dance flows smoothly, your products and your people flourish together.

Creating a Culture of Trust

Nothing erodes trust faster than chronically missed expectations. I've learned that teams perform best when working toward ambitious but achievable goals, not impossible dreams.

Promises Made and Kept

Trust in product teams isn't abstract; it's mathematical. Each promise kept builds equity; each promise broken creates debt. Over time, these small deposits and withdrawals compound into either a substantial trust reserve or a crippling deficit. Actually, the same is true in most other relationships — between countries, spouses, and business partners.

When I took over a struggling product organization, their trust account was severely overdrawn. Stakeholders had learned to double all timeline estimates, and team members were reluctant to commit to anything. Rebuilding required months of small, consistently fulfilled commitments before anyone believed our more ambitious plans.

Trust isn't just about delivering features on time. It encompasses honesty about product capabilities, transparency around challenges, and reliability in communication. Each dimension requires deliberate attention.

Honest Assessments and Realistic Expectations

Setting realistic expectations starts with honest assessments. Before committing to timelines or features, I bring key team members together for planning. If engineers say something will take six weeks, I don't arbitrarily cut it to four to please executives. Instead, I work with the team to understand what's driving the timeline and explore potential efficiencies.

I use a structured approach to these planning sessions:

1. Create psychological safety by explicitly stating that accurate estimates are more valuable than optimistic ones. "I need your real assessment, not what you think I want to hear."

2. Decompose large initiatives into smaller components where uncertainty is reduced. Estimation accuracy increases dramatically when discussing concrete, well-defined tasks rather than ambiguous features. Make sure you're using the Subtasks feature in Jira or whatever story tracking system you use.

3. Account for the full product development lifecycle, not just coding time. Design iterations, testing, documentation, and rollout planning all require allocation. Subtasks these things out to individuals and hold them accountable.

4. Incorporate known team constraints: vacation schedules, competing priorities, and learning curves for new technologies or domains. I always ask during story refinements, "Does everyone have a clear runway this month?!"

The Courage of Uncomfortable Conversations

Sometimes, setting realistic expectations means having difficult conversations with stakeholders. Years ago, I had to tell a director at Microsoft that we couldn't deliver a major feature for an upcoming TechNet conference. Rather than forcing the team into a death march, I explained why rushing would compromise quality and proposed an alternative that would still impress at the conference. He appreciated the honesty, and our team avoided burnout while delivering something solid.

These conversations are uncomfortable by nature. No one enjoys delivering disappointing news, particularly to powerful stakeholders. But I've found that senior leaders actually respect product managers who provide accurate information, even when it's not what they hoped to hear.

The key is framing these conversations productively. Rather than simply saying "we can't do that," I present options: "We can deliver feature A by your target date, or we can deliver features A, B, and C six weeks later. Here are the tradeoffs of each approach..." This shifts the discussion from binary failure to collaborative decision-making.

Underpromise & Overdeliver

One practical technique for building trust is systematically underpromising and overdelivering. By building a buffer into estimates and setting expectations slightly below what I believe we can achieve, I create space for the unexpected while allowing the team to exceed expectations regularly. I call this a "Delivery Buffer."

This doesn't mean sandbagging with ridiculous timelines. It means acknowledging reality: in complex product development, unexpected challenges always emerge. Adding a 20% buffer to thoughtful estimates isn't padding; it's pragmatism.

Alex, a product manager on one of my teams, initially resisted this approach. "Isn't it more transparent to share our actual best estimate?" he asked. Six months later, after delivering three consecutive releases ahead of schedule, he understood the value. Stakeholders were delighted, the team felt successful rather than constantly behind, and trust flourished.

The Trust Culture Reward

The ultimate benefit of a trust-based culture isn't just happier stakeholders; it's enhanced team performance. When product teams operate from a foundation of trust, they:

- Spend less energy on defensive documentation and political positioning

- Take appropriate risks without fear of disproportionate consequences

- Communicate challenges openly rather than hiding problems until they're unmanageable

- Focus on actual delivery rather than expectation management theater

The most powerful trust-building moment in my career was at a large banking institution during a critical resiliency platform launch. Technical issues with account balances emerged days before release, threatening our timeline. Rather than pressuring the team to cut corners, I announced a two-week delay. The immediate reaction was disappointment, but when we delivered a rock-solid release on the revised date, we established a reputation for reliability that served us long after.

Remember that trust isn't just a nice-to-have. It's the foundation upon which all other product leadership functions rest. Without it, even the most brilliant strategies and talented individuals will fail to reach their potential. With it, ordinary teams can achieve extraordinary outcomes through the compounding power of aligned expectations and consistent delivery.

Trust is earned in small, consistent actions; not big, flashy promises. Keep your commitments, set realistic expectations, and don't shy away from uncomfortable conversations. Underpromise, overdeliver, and always account for the unexpected. When you build trust, your team is empowered, your stakeholders are happy, and you avoid the stress of constant firefighting.

Example: LegalNexus

Dave took over as Product Director at LegalNexus, a law firm software company, when the product team was in rough shape. The previous leadership had a bad habit of promising features they couldn't deliver, leaving both attorney clients and the team frustrated.

"First day on the job, I gathered everyone together," Dave recalls. "I told them we were starting fresh with one simple rule — we only promise what we can absolutely deliver."

The team was skeptical at first. When senior executives demanded a complex e-discovery feature for an upcoming legal tech conference, Dave surprised everyone by saying no. Instead of forcing his team into a death march, he explained the quality risks and proposed an alternative solution that was achievable.

To everyone's surprise, the executives respected his honesty.

Dave introduced what he called a "delivery buffer" — adding a realistic 20% cushion to all estimates to account for unexpected challenges. This wasn't padding; it was acknowledging reality. When his team delivered three consecutive releases ahead of schedule, customers were delighted, and team morale improved dramatically.

"The magic happened about six months in," says Dave. "Suddenly our team wasn't wasting energy on defensive documentation or playing politics. They communicated challenges openly instead of hiding problems. We focused on actual delivery rather than constantly managing expectations."

The transformation became complete when a critical case management database issue emerged days before a major launch. Rather than pressuring the team to cut corners, Dave announced a two-week delay. After delivering a rock-solid release on the revised date, LegalNexus established a reputation for reliability that completely changed their position in the legal field.

"Trust isn't just a nice-to-have," Dave explains. "It's the foundation for everything else. Without it, even brilliant strategies and talented people will fail to reach their potential."

Onboarding for Future Success

Few things signal your leadership values more clearly than how you welcome new team members. Proper onboarding isn't just about administrative details; it's about transferring institutional knowledge, cultural norms, and product intuition.

I approach onboarding new product team members like teaching someone to drive. You start with the basics, let them practice in safe conditions, and gradually introduce more complexity. Throwing someone into rush hour traffic on day one is a recipe for crashes and too many leaders at too many companies do it.

For new product managers, I've developed a four-week integration program:

Week 1

Company and product immersion. Meet key stakeholders, understand business goals, and experience the product as users do.

Week 2

Team processes and tools. Learn how we make decisions, document work, and collaborate across functions.

Week 3

Shadowing active projects. Observe meetings and discussions without responsibility for outcomes.

Week 4

Taking the wheel. Lead a small, well-defined initiative with close mentorship.

Sarah joined our team as a junior product manager from a competitor. Despite her experience, our product domain was new to her. Rather than assuming her previous experience would transfer seamlessly, we went through this process methodically. By month two, she was contributing valuable perspectives from her background while successfully navigating our environment. Had we just thrown her into the deep end, we would have lost those insights as she struggled to stay afloat.

Tailoring to Individual Needs

While structure is important, effective onboarding isn't one-size-fits-all. Conversely, when we hired Alex, an experienced PM who had previously worked with several team members, we modified the approach to leverage his existing relationships while ensuring he didn't miss critical context. His onboarding focused more heavily on product and market specifics rather than team dynamics he already understood.

Create a Buddy System

One practice I've found invaluable is pairing each new hire with an onboarding buddy; an experienced team member who serves as their day-to-day guide. This person isn't their manager but a peer who can answer "stupid questions" without judgment, offer insider knowledge about team dynamics, and provide psychological safety during the vulnerable early weeks.

The buddy commits to regular check-ins, including lunch during the first week, and makes themselves available for impromptu questions. This relationship often evolves into a lasting collaboration that benefits both parties.

Documentation That Actually Helps

Many onboarding programs fail because they rely on outdated or overwhelming documentation. I try to maintain a living onboarding wiki that's reviewed every couple of months for accuracy. Rather than exhaustive detail, it provides just enough information with clear pointers to additional resources.

My teammates really like participating in these meetings and offering factoids we might be missing. This document takes a tremendous load off of them when new recruits have context and documentation to refer to. Recruits who have a few months under their belt are exceptionally valuable here. So many once confusing things are still fresh in their minds.

The most valuable component is our "Frequently Asked First Questions" section, which addresses common early confusions before they become roadblocks. This includes everything from "How do I actually get a decision approved?" to "Who should I talk to about customer data access?"

Setting Clear 30/60/90 Day Expectations

Anxiety often stems from uncertainty about expectations. I provide clear milestones for the first three months, with concrete deliverables and learning objectives for each period. These aren't performance evaluations but shared understanding of what success looks like during the onboarding phase.

For instance, by day 30, a new PM might be expected to understand the product roadmap and key metrics. By day 60, they might lead a minor feature specification. By day 90, they should be independently driving a small initiative from conception to launch.

If it becomes apparent (in many cases) that we can accelerate this rule and the new PM can accomplish a lot in shorter time, so be it. I just want them to know from the start that I don't have cruel and unrealistic expectations.

The Feedback Loop

Regular, structured feedback is essential during onboarding. I schedule brief weekly check-ins specifically focused on the onboarding experience, separate from regular one-on-ones about work. These sessions focus on identifying knowledge gaps, clarifying confusions, and adjusting the onboarding plan as needed.

Equally important is collecting feedback about the onboarding process itself. As stated earlier, each new hire helps improve the experience for future team members by highlighting what worked well and what could be better.

Good Onboarding = Good Investment

Remember: time invested in proper onboarding pays massive dividends in performance, retention, and team cohesion. A PM who ramps effectively will contribute more value over their tenure than one who struggles through a disorganized introduction.

The true measure of successful onboarding isn't how quickly someone starts producing deliverables, but how thoroughly they internalize the context, relationships, and values that enable long-term impact. Like a solid foundation for a house, proper onboarding creates stability for everything built upon it.

Polishing the Team

Product leadership is a marathon, not a sprint. Developing your team requires patience and consistent guidance over time, not erratic bursts of attention when problems arise.

This section will be longer than most. It's about all those things which make great teams, yet which don't fit under one or more clever headings. It's about connecting with your team, working out the kinks, and creating skills for handling the daily grind.

Weekly One-on-Ones

I schedule regular one-on-ones with all direct reports; 30 minutes weekly for junior team members and biweekly for seniors. These aren't status updates (we have other forums for that) but development conversations. Sometimes it's addressing specific challenges, other times discussing career goals or providing feedback.

The key is consistency. Even during our busiest periods, I make a slot for these meetings. They're an investment in the team's long-term success, not a luxury for quiet periods. Like watering plants regularly rather than drowning them occasionally, steady guidance yields better growth than sporadic intervention.

With a lot of direct reports, I see them for lunch, beers, dinner, lunches, all the time. It's customary that I get an email message the morning of a one-on-one that says, "Hey! Can we not do this today? I'm super busy. We good?" My response is usually to agree, unless there is a pressing escalation issue. I try to be accommodating and try to respect everyone's busy schedules. I try not to add to a death-by-too-many-meetings environment.

I remember a brilliant but somewhat disorganized PM struggled with some aspects of discovery and story writing. Rather than taking over his responsibilities or criticizing his methods, we used our one-on-ones to gradually build better habits. Each week, I'd offer one small, implementable suggestion, from calendar blocking to documentation templates. Over six months, the transformation was remarkable — not because of any single intervention, but through the compound interest of regular guidance.

Keeping the Team Out of Politics

Office politics are as inevitable as rain in Seattle. Your job as a leader isn't to pretend they don't exist, but to carry a real large umbrella so your team stays dry.

This should go unsaid, but let's do it. You can prevent politics within your team by fostering transparency and fair process. When decisions are made openly based on merit rather than favoritism, political maneuvering has less oxygen to breathe.

Second, shield your team from external politics while representing their interests effectively. I absorb political heat from above so my team can focus on execution, but I also ensure our work aligns with organizational priorities so we don't become irrelevant.

During a particularly tense restructuring, rumors were flying about team consolidations. Rather than letting my team spiral into anxiety, I gathered everyone for a honest conversation. I shared what I knew definitively, what was still uncertain, and my commitment to fighting for their interests. Then I encouraged everyone to channel that anxiety into excellent work, which is ultimately the best job security. I didn't bad-mouth executives who were screwing the place up and I stayed out of the negative minutia.

When political discussions do arise — whether about organizational changes or broader social issues — establish ground rules for respectful dialogue. I don't ban such conversations (that's neither possible nor desirable), but I do ensure they don't derail our core mission or create division.

Nothing can ruin a team faster than someone bitching about someone else, bringing in negativity, making stuff up, and just playing nasty. It's the dark side of politics and it's usually caused by hurt feelings, unfairness, or by a teammate feeling inadequate. I make a statement early-on with new recruits, "I play nice with others and I expect you to, as well. If you've got a problem, you're feeling hurt, or you're overwhelmed, don't complain about it, bring it to me. I want to know. I've been in your shoes."

Balancing Workloads

One of the more subtle leadership challenges is distributing work fairly across your team. Notice I said "fairly," not "equally."

Equal distribution means everyone gets the same volume of work regardless of their capacity, skills, or development needs. Fair distribution accounts for these differences while ensuring no one feels exploited.

I maintain a workload dashboard that visualizes each team member's current commitments, upcoming capacity, and recent history. This makes imbalances visible before they become problematic and facilitates transparent discussions about allocation.

When Teri, one of my strongest performers, approached me about feeling overloaded, I could immediately see she wasn't imagining things. The reports in Jira showed she had taken on 30% more features than others at her level — partly because she rarely said no and partly because stakeholders specifically requested her. We developed a re-balancing plan that included delegating some features to others and myself, saying no to new requests for a period, and coaching other team members to step up in her areas of expertise.

Remember that workload isn't just about quantity but complexity and emotional labor. Supporting a difficult stakeholder may take more energy than three projects with co-operative partners. Account for these invisible factors in your balance calculations.

Work-Life Balance for Real

The most eloquent speech about work-life balance means nothing if you're emailing at midnight and working on weekends. As a leader, you set the tone through actions more than words.

I make a point of discussing my boundaries openly. The team knows I walk my dogs every morning (meaning no 6 AM offshore meetings) and rarely respond to non-urgent messages after dinner. Rather than diminishing my authority, this transparency empowers others to establish their own boundaries.

I once had a direct report apologetically mention needing to attend his daughter's recital during a product review. I enthusiastically supported him and moved the meeting. Later, I privately reinforced that family commitments aren't something to apologize for. They're priorities we respect.

I also regularly take long walks — miles if possible. I have many meetings while I'm on those walks and I recently found out that Steve Jobs (Apple Computer founder) did the same. Most times, I ask if anyone wants to join me. During those walks, we talk a lot, get a lot of ideas and frustrations out, get to know each other, tell a few jokes, and return to the office entirely different people.

The goal isn't perfect balance every day (that's unrealistic in our field), but sustainability over time. During launch periods, we may temporarily shift into higher gear, but I ensure we downshift afterward with compensatory time off and celebration of efforts.

You may not be a drinker, but all my friends and colleagues will tell you that I sure love my wine! When I say, "Hey! Let's go out for a beer later and talk it over." It usually means you have beer (or soda, if you choose) and I'll have a nice Cabernet or Zinfandel.

The purpose is not to drink alcohol and definitely not to get drunk or be unprofessional. It's to relax in our own personal modes. While relaxing, we pull down barriers, talk candidly, and we don't feel threatened. I want to meet my team on a neutral turf — not force creative conversation in a cold, stale conference room.

Feedback and Emotional Intelligence

Honest, constructive feedback is the fuel for professional growth, but delivering it effectively requires emotional intelligence — both yours and your team's.

I use a simple method for feedback conversations: observation, impact, question. "I noticed you interrupted Jane three times during the meeting. This may have prevented us from hearing her full perspective. What's up?!"

This approach avoids accusatory language while addressing specific behaviors and their consequences. The question invites reflection rather than defensiveness.

Before giving critical feedback, I check my own emotional state. If I'm frustrated or angry, I delay the conversation until I can approach it constructively. Feedback should be about growth, not punishment or venting.

I also invest in raising the team's collective emotional intelligence. We've done workshops on communication styles, conflict resolution, and recognizing bias. These skills make feedback more effective in all directions; down, up, and across the organization.

Consider Ben, a technically brilliant product manager whose blunt communication style often created friction. Rather than just telling him to "be nicer" (unhelpful feedback), we identified specific patterns and alternatives. "When you describe ideas as 'obviously wrong,' people stop contributing. Try 'I have some concerns about that approach' instead." Combined with EQ training, these concrete suggestions helped Ben dramatically improve his interactions while maintaining his analytical strengths.

Transparency About Opportunities

Nothing breeds resentment faster than the perception of secret opportunities or hidden advancement tracks. I believe in radical transparency about growth paths, even when it occasionally means losing team members to better opportunities. I constantly promote my direct reports across organizations and a team member who leaves me (not after the first few months, of course) for a promotion and better opportunity makes me proud.

I maintain a shared document outlining skills and experiences needed for advancement at different levels. They might simply be in the form of job descriptions. When special projects or stretch assignments become available, I announce them to the entire **eligible** team rather than quietly tapping someone, unless a key stakeholder or executive make a demand for someone specific. Even then, I stop and consider the fairness.

This doesn't mean everyone gets every opportunity — merit and readiness still matter — but it does mean everyone knows what's available and what they need to demonstrate readiness.

When a junior PM expressed interest in moving toward product strategy, I connected her with our strategy team for an informational interview and helped identify aspects of her current role that could build relevant skills. Even though this might eventually lead to her leaving my team, supporting her growth was the right thing to do. She's now the CPO at an AI startup and a very good friend.

Counterintuitively, being transparent about opportunities — even external ones — often improves retention. People stay when they feel their leader genuinely cares about their growth more than just filling a role.

Example: ArtiFacts

At ArtiFacts, an artwork book publishing company in New York, their product team was experiencing friction that affected their latest art collection releases. Diana, their VP of Product, noticed the tension during her first few editorial meetings.

Elena, a brilliant senior editor, had a habit of dismissing ideas with phrases like "that won't sell" or "the art world would hate that." Team members stopped sharing concepts in meetings, and creative innovation slowed to a crawl.

Instead of publicly calling out Elena's behavior, Diana scheduled a casual coffee chat at the small café across from their SoHo office. "I noticed during yesterday's concept review you dismissed three ideas pretty quickly," she said. "When that happens, we miss out on potential innovations, and the team stops contributing. What's your take on this?"

Elena hadn't realized the impact of her communication style. She explained she was just trying to save time by eliminating options she knew wouldn't work. Diana suggested alternative phrases: "What if you asked a question instead? Something like 'I have concerns about that approach because...' opens a conversation rather than ends it."

Diana also noticed career advancement seemed mysterious at ArtiFacts. Some people received exciting artist collaborations while others, despite solid performance, seemed stuck in their roles. She created a document outlining the skills needed for each level and shared it with everyone.

When a project with a renowned photographer became available, Diana announced it to all eligible editors rather than quietly assigning it to her favorite team member. Ava, a quiet but effective editor who typically flew under the radar, expressed interest and created an exceptional book that caught leadership's attention.

Later, when Ava moved to a senior editor role on another imprint, Diana celebrated her growth publicly. "This is exactly what good publishing leadership looks like," she told the team. The result? Applications to join Diana's team doubled, and team members became more open about their career aspirations, knowing Diana would support their growth.

Merit Over Familiarity or Cronyism

Promotion decisions reveal your true values as a leader. Do you reward loyalty or impact? Visibility or value? Politics or performance?

I establish clear, documented criteria for promotion at each level, focusing on observable behaviors and outcomes rather than subjective assessments or time served. These criteria are shared with the team before evaluation cycles so everyone understands the bar.

When considering promotions, I gather input from multiple stakeholders to overcome my own biases. It's natural to favor people whose working styles mirror your own or who make your life easier, but that's not necessarily who delivers the most value.

For one of my teams, David had been around a bit longer than Leila, and we had an excellent rapport. He was good friend and when a senior position opened, my instinct leaned toward David, of course. However, when I evaluated both against our promotion criteria and collected feedback, Leila's impact was much stronger. Promoting her despite my personal comfort with David sent an important message about what our team truly values. I then explained all that to David over dinner and he, too, is now a leader practicing the same type of evaluations. It lent itself to his growth, as well. We speak often and he's a stellar director at a large Charlotte area investment banking firm.

For those not promoted, I provide specific, actionable feedback about the gap between current performance and the next level, along with a development plan to bridge that gap. David is proof that this transforms disappointment into motivation rather than resentment.

Recognition & Team Culture

Recognition costs nothing but yields enormous returns in motivation and loyalty. Yet many leaders underutilize this powerful tool or apply it inconsistently.

I build recognition into our rhythms. Team meetings always include time for highlighting exceptional work or extra effort. These aren't vague "good job" statements but specific acknowledgments: "Lisa's user research protocol uncovered insights we would have missed, fundamentally improving our approach."

More importantly, I ensure recognition reaches the right people. When executives compliment a project, I make sure they know who specifically made it successful, especially those in less visible roles. I forward praise emails, mention contributors in leadership meetings, and create opportunities for team members to present their own work.

Alex, an introverted but brilliant BA (Business Analyst) on on of my teams, had transformed our understanding of a key banking customer segment through methodical work over months. Rather than summarizing his findings myself at our next planning meeting, I coached him on presenting them and gave him the spotlight. The recognition from senior leaders boosted his confidence more than any praise from me.

Remember that people have different recognition preferences. Some value public acknowledgment, others prefer private appreciation, and still others care most about growth opportunities as recognition. Get to know these preferences through conversation and observation.

I have a small checklist of monthly things to get done — when I have a minute to breathe. One of those items is named "Send Kudos!"

Compassion & Being Human

Product development is intense, deadline-driven work. Amid this pressure, it's easy to forget that we're leading humans, not resources.

People bring their whole selves to work, including the challenges they face outside the office. Medical issues, family crises, mental health struggles; these don't disappear during working hours. Show some freakin' compassion, already!

I've found that compassion and accommodation during personal challenges earns loyalty that money just can't buy.

When Kevin's partner was diagnosed with a serious illness, we restructured his workload and created flexible arrangements so he could attend medical appointments. The team stepped up to support him, and when he returned to full capacity months later, his commitment to the team was stronger than ever. We were so happy to have his smile back in the office.

Compassion isn't just for crises. It means recognizing that parents might need different schedules, that some people have energy peaks at different times of day, that religious observances matter, and that mental health needs aren't weakness but human reality.

The key is consistency. Compassion shouldn't be reserved for your favorite team members or those in certain roles. Everyone deserves the same human consideration, though the specific accommodations may differ based on individual needs.

When someone on my team is having a hard go of it, I try to circulate a greeting card, have everyone sign it (with personalized remarks), and I attach it to a delivery of chocolate or their favorite food. Because of these types of gestures, my team members usually cannot wait to get back to work. That's the place they go where they're respected, cared about, and where "work" turns into a "gift."

Mandatory Recharging

In our always-connected culture, truly disconnecting has become almost taboo. As a leader, I actively counter this unhealthy norm by modeling and encouraging genuine time off.

I take my vacation time fully and visibly. Before leaving, I announce that I'll be truly offline (not "checking in occasionally") and demonstrate trust by delegating authority clearly rather than maintaining a supervision chain.

When team members are on vacation, I have a strict policy against contacting them except for genuine emergencies (which are extremely rare if you've planned properly). If someone regularly responds to messages while away, I follow up with them; not to thank them for their dedication, but to reinforce that disconnecting is part of our culture.

After returning from time off, I share my experience openly, normalizing the benefits of disconnection. "That two-week break gave me perspective on our prioritization that I couldn't see when immersed in daily details."

Jenny, an outstanding but somewhat intense product manager, initially took pride in never fully disconnecting. After several burnout warning signs, I insisted on a completely offline vacation. The difference in her energy and creativity upon return was remarkable — and a powerful example for the team of why true breaks matter.

Remember that time away includes mental health days, not just planned vacations. I make it clear that occasional "recharge days" are legitimate uses of personal time, not something to be embarrassed about.

Fighting for Your Team

Perhaps the most important role of a product leader is being the team's champion within the broader organization. This means advocating for resources, recognition, compensation, and reasonable expectations.

Effective advocacy requires building political capital through reliable delivery and strategic alignment, then spending that capital judiciously on behalf of your team. I make sure our work visibly connects to company priorities so when I need to advocate, I'm speaking a language leadership values.

When budget constraints threatened to eliminate two positions from my team during a reorganization, I came prepared with data showing the specific business impact these roles had delivered and the risks of their elimination. Rather than just emotionally arguing for "my people," I made a business case that resonated with decision-makers.

Sometimes advocacy means pushing back on unreasonable deadlines or scope changes, even when they come from powerful stakeholders. I frame these conversations around shared goals: "To deliver the quality we all want, we need either more time or reduced scope. Let's discuss which tradeoff better serves our ultimate objective."

Your team watches how you represent their interests when they're not in the room. When they see you fighting fairly but firmly on their behalf, it builds the kind of loyalty that drives exceptional performance.

Your Leadership Legacy

Leadership isn't a destination but a journey of continuous growth and adaptation. The true measure of your success isn't just the products you ship, but the people you develop.

Years after leading any particular team, your legacy won't be feature releases or OKRs (Objectives and Key Results) achieved. It will be the careers you advanced, the skills you nurtured, and the culture you built. Former team members carrying your values into new organizations is perhaps the highest form of product leadership impact.

Remember that everything you do as a leader sets precedent and shapes culture. Your responses to pressure, your treatment of mistakes, your allocation of opportunities; all these small daily choices accumulate into the team environment.

Lead with authenticity, integrity, and genuine care for both your products and your people. Technical skills might make you a competent manager, but these human qualities make you a leader worth following.

Product leadership at its best isn't about control but enablement; creating the conditions where talented people can do their best work together. Master this art, and there's no product you cannot build.

Concepts in Action: Sorensen Technologies

A software company in Seattle called Sorensen Technologies was facing a critical crossroads. After five years of promising growth, the product development teams had begun to fracture under the weight of their own success. What had once been a tight-knit group of passionate innovators had evolved into siloed teams plagued by missed deadlines, political infighting, and diminishing trust.

Enter Sharon Paulson, newly appointed Director of Product who had been brought in to transform the chaotic symphony into a harmonious orchestra. Like a wise conductor who knows every instrument has its role, Sharon set about implementing the principles that would breathe new life into Sorensen.

Lane Keeping & Team Hierarchy

Sharon's first order of business was establishing clear lanes for each role. "Product managers focus on the 'what' and 'why,' designers on the 'how it works and feels,' engineers on the 'how it's built,' and researchers on 'what users need,'" she explained during her first all-hands meeting.

She noticed that Darius, a talented but overzealous product manager, had a habit of designing interfaces during meetings, effectively stepping on the design team's toes. This had created resentment among designers who gradually stopped bringing their best ideas forward.

Rather than publicly criticizing Darius, Sharon clarified that his primary focus should be articulating problems clearly and setting success metrics, while leaving solution exploration to designers. She emphasized that Darius could still contribute ideas, but within a collaborative method rather than prescriptively.

Within weeks, the design team began sharing more innovative concepts, knowing their expertise was respected.

Supporting Without Undermining

Sharon adopted a "disagree and commit" mindset with her senior product managers. When Priya, an experienced SPM, made a decision about feature prioritization that Sharon privately disagreed with, she chose to support it publicly rather than override it. The decision wasn't catastrophic, and Sharon recognized that preserving Priya's authority and creating a learning opportunity outweighed the marginal benefit of what Sharon considered an "optimal" decision.

The result? Priya's team executed flawlessly, and the feature performed better than expected precisely because Priya had deeper domain knowledge about their specific users.

Creating a Culture of Trust

Trust at Sorensen had been severely damaged before Sharon's arrival. Previous leaders had consistently overpromised and underdelivered, causing stakeholders to habitually double timeline estimates. Sharon's approach was mathematical: each promise kept would build equity; each promise broken would create debt.

She instituted planning sessions where she explicitly stated, "I need your real assessment, not what you think I want to hear." When engineers said something would take six weeks, she didn't arbitrarily cut it to four weeks to please executives.

This was put to the test when the marketing team demanded their new analytics dashboard be ready for an upcoming trade show. Rather than forcing the team into a death march, Sharon had the courage to have an uncomfortable conversation with the Chief Marketing Officer. Instead of simply saying "we can't do that," she presented options: "We can deliver the core metrics by your target date, or we can deliver the complete dashboard with all visualizations three weeks later."

The CMO appreciated the honesty and chose the first option. The team delivered exactly what they promised, on time, beginning to rebuild the trust account.

Onboarding for Future Success

When Sorensen Technologies hired Li, a brilliant developer-turned-product manager, Sharon didn't throw him into the deep end. Instead, she implemented her four-week integration program: Week 1 focused on company and product immersion; Week 2 on team processes and tools; Week 3 on shadowing active projects; and Week 4 on leading a small, well-defined initiative with close mentorship.

She paired Li with an onboarding buddy, Jamie, an experienced PM who served as his day-to-day guide – someone who could answer "stupid questions" without judgment and provide psychological safety during the vulnerable early weeks.

Sharon also provided clear 30/60/90 day expectations. By day 30, Li was expected to understand the product roadmap and key metrics. By day 60, he would lead a minor feature specification. By day 90, he should be independently driving a small initiative from conception to launch.

This methodical approach paid dividends as Li became a productive team member in half the time it had previously taken new hires to get up to speed.

Polishing the Team

Sharon scheduled regular one-on-ones with all direct reports: 30 minutes weekly for junior team members, biweekly for seniors. These weren't status updates but development conversations focused on growth opportunities and addressing challenges.

She also created a workload dashboard that visualized each team member's current commitments, upcoming capacity, and recent history. This made imbalances visible before they became problematic. When the dashboard revealed that Aaron had taken on significantly more projects than others at his level, Sharon implemented a rebalancing plan that included delegating some features to others and temporarily saying no to new requests.

Sharon led by example when it came to work-life balance. She made a point of discussing her boundaries openly. The team knew she took walks every afternoon and rarely responded to non-urgent messages after dinner. Rather than diminishing her authority, this transparency empowered others to establish their own boundaries.

Recognition became built into the rhythm of the organization. Team meetings always included time for highlighting exceptional work with specific acknowledgments: "Melissa's user research protocol uncovered insights we would have missed, fundamentally improving our approach to the mobile navigation."

The Results

Within a year, Sorensen had transformed. Product releases were hitting their deadlines with fewer bugs. Cross-functional collaboration flourished as each role felt respected for their expertise. Employee retention improved dramatically, and when asked why they stayed, team members frequently cited the supportive leadership and culture of trust.

The true measure of Sharon's success wasn't just the improved products or met OKRs; it was the people she developed. When Priya eventually left to become a VP of Product at another company, she implemented many of the same principles Sharon had taught her. Similarly, when Li was promoted to lead a new team within Sorensen, he carried forward the onboarding approach that had made such a difference in his own experience.

Through thoughtful implementation of leadership principles — from lane keeping to trust building, from proper onboarding to genuinely caring for her team — Sharon had turned Sorensen Technologies from a mess of competing roles into a well disciplined team, proving that great product management is as much about orchestrating people as it is about building products.

Execution Essentials

Lane Keeping & Team Hierarchy

- Clearly define roles: Product managers handle "what" and "why," designers manage "how it works and feels," engineers determine "how it's built."

- Adopt a "disagree and commit" mindset with your direct reports — support their decisions even when you'd choose differently.

- Push decision-making to the lowest appropriate level while providing support structures.

- Document decision rights using RACI (Responsible, Accountable, Consulted, Informed) methods.

- Create permeable boundaries; encourage cross-functional understanding without role confusion.

Creating a Culture of Trust

- Build trust mathematically: each promise kept is a deposit, each broken commitment is a withdrawal.

- Underpromise and overdeliver by adding a 20% buffer to estimates. It's not padding, it's pragmatism.

- Have the courage for uncomfortable conversations with stakeholders. Present options rather than simply saying "no."

- During planning sessions, explicitly state, "I need your real assessment, not what you think I want to hear."

- Focus on honest assessments. Don't arbitrarily cut timeline estimates to please executives.

Onboarding for Future Success

- Implement a four-week integration program: product immersion, team processes, shadowing, and a small initiative.

- Create a buddy system – pair new hires with experienced team members who can answer "stupid questions" without judgment.

- Maintain a living onboarding wiki with "Frequently Asked First Questions" that's reviewed regularly.

- Set clear 30/60/90 day expectations with concrete deliverables and learning objectives.

- Schedule brief weekly check-ins specifically focused on the onboarding experience, separate from regular one-on-ones.

Supporting Team Development

- Schedule regular one-on-ones focused on development, not status updates: 30 minutes weekly for juniors, biweekly for seniors.

- Ask questions before offering solutions: "What factors did you consider?" rather than "Here's what you should do."

- Use a simple method for feedback conversations: observation, impact, question.

- Establish clear, documented criteria for promotion at each level, focusing on observable behaviors and outcomes.

- Gather input from multiple stakeholders to overcome your own biases when considering promotions.

Work-Life Balance & Team Care

- Lead by example! Set and respect your own boundaries visibly through actions, not just words.

- Create a workload dashboard to visualize each team member's commitments and prevent imbalances.

- Recognize that workload isn't just about quantity but complexity and emotional labor. Account for these invisible factors.

- Build recognition into your rhythms; including time for highlighting exceptional work in team meetings.

- Actively encourage true disconnection during vacations; no "checking in occasionally."

Leadership Effectiveness

- Shield your team from external politics while representing their interests effectively.

- Fight for your team with data; making business cases that resonate with decision-makers.

- Show compassion during personal challenges; creating flexible arrangements when needed.

- Maintain transparency about growth paths and opportunities, even when it means potentially losing team members.

- Remember that your leadership legacy is the careers you advanced, the skills you nurtured, and the culture you built.

The Role

Chapter 5

Stakeholder Management

I've built and managed thousands of features, and somehow survived exponentially more stakeholder meetings. If there's one thing I've learned, it's that your product's success depends as much on how you manage people as how you manage the product itself.

Stakeholder management isn't just a nice-to-have skill. It's the invisible force that either propels your product forward or anchors it to an eternal state of mediocrity. This chapter explores the nuances of this critical discipline that separates legendary product leaders from the ones who are "looking for new opportunities" on LinkedIn.

Define, Relate, Communicate

At its core, stakeholder management follows a simple formula: **Define** who matters, **Relate** to them authentically, and **Communicate** deliberately. Miss any of these elements, and you're building a house of cards.

Defining Stakeholders

While building an online banking payment feature in 2021, I assumed my team had all my stakeholders identified: executive sponsors, Legal, and of course, the dev team. Two weeks after launch, our banking team threw an urgent meeting on my calendar, fuming about the flood of customer complaints their team couldn't handle. Somehow, we had forgotten to include Banking Support — the people who would be on the front lines dealing with customer confusion.

Stakeholder Segments

Stakeholders are anyone with an interest in or influence over your product. They include:

- Internal business units: Engineering, marketing, sales, support, operations
- Leadership: Executives who approve budgets and expect results
- Customers: The people who will actually use your product
- Partners: Third-party vendors, integration partners, and distribution channels
- External entities: Regulators, industry groups, even competitors

The first step in stakeholder management is creating an exhaustive list. Think beyond the obvious. For every feature or product, ask yourself: "Who will care about this? Who will it impact? Who has influence over its success?" Don't wait for stakeholders to find you! By then, the house is on fire and it's usually too late.

Hidden Influencers

Some of your most important stakeholders won't appear on any org chart or in your customer database. I call these the "invisible stakeholders"; people and entities with outsized influence that often go overlooked.

For instance, if you're going to build something which has a lot of executive eyeballs on it, you might want to include those executives' assistants. You can derail your launch, if the gatekeepers of executive communication don't relay messaging to the bosses. These invisible power players will suddenly become pivotal.

Other invisible stakeholders might include:

- The maintenance staff who will need to integrate your hardware product
- Administrative assistants who often make purchasing decisions
- IT security teams who can veto your product on compliance grounds
- Industry analysts whose opinions shape market perception

These invisible stakeholders deserve special empathy. They rarely get the glory but often shoulder the burden of your product's implementation. Identify them early, and you'll have powerful allies when you need them most.

Stakeholder Mapping

Once you've identified your stakeholders, it's time to map them according to their power and interest. This classic approach helps prioritize your limited attention span, which — if you're like me — has been irreparably damaged by years of emergency product meetings and far too many Skype notifications.

For smaller, less interesting feature updates, I simply make notes in my Jira issue comments and/or tag other users. For more visible and important updates, I create a simple two-axis grid (sometimes in a spreadsheet). This grid will include 4 types of stakeholders:

High Power, High Interest

These are your key players. They care deeply about your product and have the authority to make or break it. Treat them like the VIPs they are.

High Power, Low Interest

These are your sleeping giants. They don't particularly care about your product until something goes wrong — then they can squash it like a bug. Keep them satisfied but don't overwhelm them with details.

Low Power, High Interest

These are your cheerleaders. They're passionate about your product but don't necessarily have direct authority. Keep them informed and leverage their enthusiasm.

Low Power, Low Interest

These are your bystanders. Monitor them, but don't lose sleep over them.

For example, when launching your payment processing system, keep in mind that your CFO is clearly a high-power, high-interest stakeholder who needs deep engagement. The marketing team for an unrelated product line is a low-power, low-interest group that simply needs occasional updates.

This mapping isn't just an academic exercise. It's your battle plan. It determines how you'll allocate your scarcest resource: your attention.

Building Stakeholder Relationships

Now comes the part that makes many product managers uncomfortable: building actual human relationships. I've known brilliant product minds who could architect complex systems but couldn't navigate a conversation with the head of sales. Their products usually died quiet deaths, mourned only by the product team that loved them.

Long vs. Short-term Stakeholders

Stakeholder relationships come in two varieties: strategic long-term partnerships and tactical short-term alignments.

Strategic stakeholders are your perennial collaborators. These relationships transcend individual projects and form the foundation of your influence network. For most, building rapport with C-suite leaders early in their career pays dividends for years, giving them an ally when the value of some features heats up. Tactical stakeholders matter intensely for specific initiatives but may fade into the background afterward.

The trick is recognizing which is which and investing accordingly. I've wasted months cultivating relationships with stakeholders who became irrelevant after a launch, and I've underinvested in relationships that could have eased years of product development.

Stakeholder Empathy

Behind every stakeholder title is a human with hopes, fears, and a performance review hanging over their head. Understanding their personal motivations is often more important than understanding their department's official position.

Once, when our marketing director pushed back on a product simplification I proposed, her stated reason was "brand consistency." After a candid conversation over a few beers, I learned her real concern: she'd just pitched a major campaign based on the existing feature set and feared looking unprepared to her new boss. Aha! There was more to it.

By addressing her actual need (saving face with leadership) rather than her stated objection (brand consistency), we found a compromise that worked for both of us. Sometimes product management feels more like being a therapist than a technologist — except therapists probably have better work-life balance and don't get pulled into weekend BCP (Business Continuity Planning) emergencies.

Stakeholders are Not All Equal

Here's a hard truth that took me years to accept: not all stakeholders deserve equal time, attention, or influence. Playing politics makes many of us uncomfortable, but ignoring the reality of organizational power dynamics is a fast track to product failure.

I used to pride myself on giving everyone "equal voice" in product decisions. This democratic approach earned me goodwill but produced watered-down products that tried to please everyone and delighted no one.

Now I'm more strategic. I analyze which stakeholders have the most relevant expertise for specific decisions and which have the organizational clout to drive or block implementation. These stakeholders get more of my attention. It's not about favoritism; it's about effectiveness.

For instance, when determining pricing strategy, I'll give more weight to the input from sales and finance than from engineering. When making technical architecture decisions, the reverse is true. This isn't dismissing anyone's value. It's simply recognizing where different expertise matters most.

Favor Group Engagement Over One-on-Ones

If you're scheduling separate meetings with each stakeholder, you're not just wasting time — you're actively creating problems. I learned this lesson painfully after watching my carefully cultivated individual stakeholder agreements collapse spectacularly in our first group review.

Individual meetings create:

- Information silos where each stakeholder believes their priorities reign supreme

- Hidden conflicts that only emerge when stakeholders finally interact

- Exponentially increasing communication overhead as your stakeholder count grows

Stop it.

Instead, bring stakeholders together early and often. Yes, it's messier. Yes, conflicts will emerge. But addressing these dynamics in the open allows you to facilitate real resolution rather than managing a complex web of separate agreements that inevitably contradict each other.

For one project, I replaced twelve separate stakeholder meetings with three cross-functional sessions. Not only did this save about 20 hours of meeting time, but it forced important trade-off discussions to happen with all perspectives represented.

Those who didn't show up were automatically giving proxy to those who did. Sorry!

Don't Go It Alone!

A common mistake I see with new product leaders is making themselves the sole point of contact for all stakeholder interactions. This is stupid and creates an unsustainable bottleneck and misses valuable development opportunities for your team.

Instead, distribute stakeholder relationships across your product team. This serves multiple purposes:

- It scales your stakeholder management capacity

- It develops your team members' organizational savvy and leadership skills

- It creates redundancy so relationships don't collapse when someone takes vacation or leaves

When I took over a large banking resiliency platform, I inherited a stakeholder network entirely dependent on my predecessor. After she went onto another product, we lost weeks while I rebuilt those connections. Now I ensure each key stakeholder relationship has at least two team members involved, if possible.

For instance, your product manager focusing on the analytics feature owns the relationship with the data science team, while you maintain a secondary connection through quarterly check-ins. If either of you leaves, the relationship with your team remains intact.

Example: TrendMobile

TrendMobile, an Atlanta tech company, had just begun work on their flagship product; an auto parts sourcing app called "PartsFinder." As their newly hired Director of Product, Tara faced the challenge of managing diverse stakeholders across the organization.

Tara recognized the difference between strategic long-term and tactical short-term stakeholders. She invested heavily in building relationships with the CFO and CTO who would influence product decisions for years, while maintaining lighter connections with temporary contractors working on the initial UI design.

During an early planning meeting, the marketing director, Jake, seemed oddly resistant to the app's proposed search functionality. Rather than accepting his stated reason about "brand consistency," Tara invited him for coffee where she discovered his real concern: he'd already promised major auto parts retailers a different search experience. By addressing his actual fear (damaged retailer relationships) rather than his stated objection, they found a compromise that worked for both teams.

Tara was careful about stakeholder equality. When determining pricing models for premium features, she gave more weight to finance and sales input. For inventory categorization decisions, she prioritized feedback from the supply chain team and auto parts specialists. This wasn't playing favorites; it was recognizing where different expertise mattered most.

Instead of holding separate meetings with each department head, Tara instituted cross-functional working sessions every two weeks. This approach not only saved countless hours but forced important trade-off discussions to happen with everyone present. Stakeholders who missed these sessions understood they were delegating their voice to others.

Tara also distributed stakeholder relationships across her team. Her junior product manager Raj owned the relationship with the engineering team, while her senior analyst Meghan maintained connections with the auto parts suppliers. When Raj eventually left for another opportunity, the engineering relationship remained intact through Meghan, preventing any disruption to the product development process.

> This thoughtful approach to building stakeholder relationships helped TrendMobile launch PartsFinder three weeks ahead of schedule with enthusiastic support from across the company.

Stakeholder Communication

With stakeholders defined and relationships established, communication becomes your primary tool for ongoing management. The best product strategy in the world is worthless if it's stuck inside your head — or worse, misunderstood by those who need to execute it.

Early, Constant, and Transparent

The cardinal sin of stakeholder communication is the surprise. Nothing undermines trust faster than stakeholders learning about significant decisions after they're made or, even worse, from someone outside your team.

Early communication means involving stakeholders at the inception of ideas, not after plans are fully baked. I've learned to share even half-formed concepts with key stakeholders, making it clear I'm seeking input, not presenting final decisions.

Constant communication maintains alignment through the inevitable twists and turns of product development. Establish regular touchpoints that don't depend on crises to trigger interaction.

Transparent communication builds trust even when delivering unwelcome news. When I once had to delay a major feature release due to technical debt issues, I shared the complete situation with stakeholders rather than offering vague excuses. This temporary pain preserved long-term credibility.

How Best to Communicate with Stakeholders

Effective stakeholder communication uses multiple channels, each suited to different purposes:

Pulse Surveys

Quick temperature checks that give stakeholders a voice without requiring meetings. I sometimes use quarterly pulse surveys to gauge satisfaction with product direction and communication effectiveness.

Newsletters

Regular updates that maintain awareness without requiring immediate action. A monthly product newsletter might include upcoming releases, recent wins, and recognition of cross-functional contributions.

Email Updates

Targeted communications for specific developments or decisions. If you're shifting your API strategy, send a dedicated email explaining the rationale and implications.

Collaborative Workspaces

Shared environments where stakeholders can track progress and provide input asynchronously — my favorite. For small companies, use Slack for collaboration. For larger enterprises, use Confluence or Notion and create a site for your team to keep others updated on their terms.

Meetings

Video catchups and in-person meetings with key stakeholders is sometimes necessary, but it shouldn't be the norm. That leads to meeting fatigue and usually wastes the most time. Keep the meeting short, always provide an itinerary in the invitation, and invite only key players.

Show-and-Tells

Live demonstrations that make progress tangible. Our bi-weekly demos give stakeholders concrete evidence of advancement and opportunities to provide feedback.

Now, I know I just pooh-poohed meetings (above). However, this is different. Show-and-tells can be controlled. They are demonstrations of things which have gotten done and a forum for constructive feedback.

Keep the Medium Smart

The best stakeholder communicators match the medium to the message and the audience. A technical architecture change might warrant a detailed document for the engineering team but a short executive summary for the C-suite.

Don't brown-nose and invite the CTO and CEO to a Zoom meeting, if it's not necessary. They're probably not going to show-up, or they will show-up and enter the call right at the time when others are complaining about the recent restructuring.

Tailoring Messages to Stakeholder Segments

Different stakeholders care about different aspects of your product. Customizing your communication to address their specific concerns demonstrates respect for their time and perspectives.

For executives, focus on business outcomes, market positioning, and financial implications. Our CEO cares deeply about how our products differentiate us from competitors but has little interest in implementation details.

For technical teams, emphasize architectural decisions, technical dependencies, and implementation considerations. For example, your engineering leads need to understand how product choices affect their technology roadmap.

For customer-facing teams, highlight user benefits, potential objections, and competitive comparisons. Your sales team needs to understand how to position new features against market alternatives.

This tailoring doesn't mean creating entirely separate narratives. The core story should remain consistent. It's about emphasizing the elements most relevant to each audience.

Often, if I'm sending out an email update on a particular feature set, I'll make nice colorful headers for certain sections. The C-suite skims for whats of interest to them, as do all other interested stakeholders.

Another clever thing I like to do is color key statements, which are buried in paragraphs, with orange or red. That allows busy colleagues to skim for just those items, being fully updated in seconds. It's kind of like pull-quotes in longer articles you might see in magazines.

Getting Stakeholders On Board

Stakeholder alignment exists on a spectrum from awareness to active championship. Knowing what level of buy-in you need from each stakeholder prevents both under-engagement (where stakeholders block you later) and over-engagement (where you waste time securing unnecessary approvals).

For major initiatives (only), I use a simple RACI grid:

- **Responsible:** Who will do the work?
- **Accountable:** Who will make final decisions?
- **Consulted:** Who must provide input before decisions are made?
- **Informed:** Who must be kept updated but doesn't need to be consulted?

Most importantly, I make sure this RACI is shared with some or all of the stakeholders, if possible, and if it doesn't entail any privacy or security implications. This clarity prevents the all-too-common scenario where someone thought they had decision rights they didn't actually possess.

For formal sign-off processes, be explicit about what stakeholders are approving. Are they endorsing the overall direction, the specific implementation approach, or the detailed requirements? Ambiguity here leads to misaligned expectations and painful backtracking.

Feature / Stakeholder	Ben Product	April Design	Shea Engineering	Leandra Content	Travis Sales	Brian APIs
Splash Image	R	R	A	C		
Login Form	R	A	R	I		C
Accounts Screen	R	R	A	A	I	A
Products List	A	A	R	I	C	A
Discounts Feature	A	I	A		R	

A very simple example of a RACI for possibly a banking app.

Keeping Stakeholders in the Loop

Once initial buy-in is secured, maintaining appropriate information flow prevents stakeholders from becoming disconnected or, worse, actively concerned about your product's progress.

The key word here is "appropriate." Flooding stakeholders with every detail creates information fatigue and trains them to ignore your communications. Providing too little information breeds uncertainty and rumor.

I follow a simple rule: communicate at the frequency and detail level that gives stakeholders confidence without creating burden. For most stakeholders, this means regular high-level updates with the option to drill down deeper if desired.

A good tiered approach might be:

- Weekly updates for directly involved teams
- Bi-weekly summaries for adjacent stakeholders
- Monthly highlights for executive sponsors
- Quarterly reviews for distant but important stakeholders

Each level distills information appropriately while maintaining a consistent narrative across audiences.

Example: OrbitalCRM

OrbitalCRM, a mid-sized software company specializing in customer relationship management solutions, recently began developing a new mobile application called "MobiSync" to help small businesses manage client interactions on the go. The product leader, Rebecca Chen, implemented a structured stakeholder communication plan based on appropriate information flow principles.

Rebecca recognized that different stakeholders required different levels of detail and frequency of updates. Instead of bombarding everyone with the same information, she created a tiered communication approach that gave stakeholders confidence without creating unnecessary burden.

Her plan looked like this:

For the development and design teams directly building MobiSync, Rebecca scheduled concise 15-minute weekly updates. These meetings included detailed progress reports, technical challenges, and immediate next steps. Additionally, she maintained a dedicated Slack channel where daily achievements and blockers were shared.

For the marketing, sales, and customer support teams, Rebecca organized bi-weekly summary emails highlighting key milestones, upcoming features, and potential market positioning. These departments needed to prepare for the launch but didn't require intricate development details.

OrbitalCRM's executive team received monthly highlight reports focusing on business outcomes, market positioning, and financial implications. Rebecca created a one-page dashboard with green/yellow/red indicators for budget, timeline, and feature completion; allowing executives to quickly gauge project health.

For external partners and distant stakeholders, such as the payment processing vendor and industry analysts, Rebecca scheduled quarterly review sessions. These comprehensive but streamlined presentations provided the big picture without overwhelming these stakeholders with information they didn't need regularly.

When an unexpected security vulnerability was discovered midway through development, Rebecca's communication system proved its worth. Rather than sending a panic-inducing mass email to all stakeholders, she tailored her communications:

- The development team received detailed technical information about the vulnerability and potential solutions through their daily channel and an emergency meeting.

- Adjacent stakeholders got a bi-weekly update that acknowledged the challenge but emphasized the team was implementing a solution that wouldn't impact the timeline.

- Executives received their monthly report with a yellow indicator for the security component, a brief explanation of the issue, and assurance that mitigation was underway.

- Quarterly partners weren't notified until the next scheduled update when Rebecca could report both the challenge and its successful resolution.

This approach prevented information fatigue while maintaining appropriate transparency. Each stakeholder group received exactly what they needed: technical teams got details for action, adjacent teams received context for planning, executives got strategic implications, and distant stakeholders got relevant summaries.

> By the time MobiSync launched, stakeholders at every level felt appropriately informed and engaged; neither overwhelmed with minutiae nor left in the dark about important developments.

Learning to Say No

Product leaders rarely have direct authority over most of their stakeholders. We can't order the sales team to prioritize our feature or demand that Legal expedite their review. Our success depends on influence; the ability to drive action without formal power.

The worst stakeholder managers are human vending machines, dispensing whatever feature is requested with the push of a button. The best are thought partners who sometimes push back, except with constructive alternatives.

Early in my career, I said yes to nearly everything, creating a bloated product and an unsustainable roadmap. Now I've learned that "no" can be the beginning of a conversation rather than the end of one.

If your sales leader is requesting a complex customization feature that would fragment your codebase, don't simply refuse. Instead, acknowledge the underlying need (serving a key customer segment) and propose an alternative approach using configuration rather than customization. This "yes to the problem, no to the specific solution" approach maintains the relationship while protecting product integrity.

Compromise and Influence

Stakeholder management often comes down to negotiation. Rather than viewing this as a zero-sum game, look for creative compromises that address core needs while maintaining product coherence.

When negotiating with stakeholders, I focus on interests rather than positions. Instead of debating their specific feature request, I dig into what problem they're trying to solve or what opportunity they see.

For example, if your support team is requesting a complex admin interface that would delay your release by months, explore their underlying need will reveal they're really wanting to reduce ticket resolution time. Find a much simpler solution that addresses this need without the extensive development effort.

Potential Stakeholder Pitfalls

Oh, you think that's all there is to managing stakeholders? Well, once you get the hang of it, there are a few things you should keep an eye out for.

Don't Let Stakeholders Plan Your Features

There's a fine line between incorporating stakeholder input and letting stakeholders design your product. The former is essential; the latter is abdicating your responsibility as a product leader.

Stakeholders should absolutely inform you about problems, opportunities, constraints, and contexts. They should not be dictating specific features, interfaces, or implementations. That's your job.

When stakeholders come with fully formed feature ideas, thank them for their input, then reframe the conversation around the underlying needs. "That's an interesting approach. Help me understand what problem this would solve for you." This steers the discussion back to territory where their expertise is most valuable; defining the what and why, not the how.

Never Flow Requests Directly Into Your Dev Team

One of the quickest ways to undermine your product discipline is becoming a requirements pass-through, relaying stakeholder requests directly to your development team without critical evaluation.

I've seen product managers who essentially function as administrative assistants to their stakeholders, diligently logging every request and passing it to development without adding value. This not only clutters your roadmap but damages your credibility with the development team.

When I was consulting for a large insurance company, a particular VP would be on many of his team's scrum calls. All of a sudden, he would burst in with something like, "Here's how this has to work...(insert unvetted bullshit feature here)...James, please add a story on that, tag me to it, and let me know when it's done."

It used to piss me off to no end! And it happens a lot. Poorly trained executives can sometimes act like the Head Feature Maker. I eventually put a stop to it and the team thanked me profusely, let alone started getting so much more done for the customer.

Every stakeholder request should go through your product filter: Does this align with our strategy? Does it address a validated user need? Is it the most efficient way to solve the problem? Is it worth the opportunity cost? Only after this analysis should requirements be shaped and prioritized for development.

If an executive is taking over your job, make sure you keep him or her in their lane. Have a meeting with them and simply state, "Listen. I know you're trying to help here and I'm grateful, but this will work best if you tell me the outcome you're looking for in the big picture, and I go and get that done for you and with the team?"

Defining Success on Your Terms

Similarly, don't let stakeholders define what success looks like for your product. While their input is valuable, ultimate success metrics should align with your product strategy and organizational goals.

When I once launched a customer portal, the sales team wanted to measure success by how many new clients it helped them sign. This was a relevant metric but made our success dependent on their execution. Instead, we established primary success metrics around user engagement and task completion, with sales impact as a secondary consideration.

Be Transparent About Mistakes

Nothing ruins stakeholder trust faster than obscuring problems or, worse, attempting to rewrite history. When mistakes happen — and they surely will — transparent acknowledgment is your best path forward.

After a particularly painful release where my team missed several committed features, I was tempted to focus only on what we delivered. Instead, I explicitly acknowledged the misses, explained what went wrong, and outlined our corrective measures. Rather than diminishing confidence, this honesty actually strengthened our stakeholder relationships by demonstrating accountability.

This transparency has limits, of course. Which leads us to...

Discretion With Sensitive Information

While transparency builds trust, discretion preserves it. Not all information should be shared with all stakeholders, particularly when it comes to confidential business data, personnel issues, or competitive intelligence.

I've seen product leaders damage relationships by sharing information from one stakeholder with another inappropriately. This may seem helpful in the moment but inevitably backfires when trust is broken.

Maintain appropriate information boundaries. When stakeholders ask for information you can't share, explain the constraint rather than being evasive. Most stakeholders will respect clear boundaries more than they'll appreciate inappropriate sharing that might later be turned against them.

Stakeholder Management Tools

Modern stakeholder management benefits from purpose-built tools that systematize what once relied entirely on interpersonal skills and memory. These tools don't replace the human element but can significantly enhance your effectiveness and scalability.

Simply Stakeholders: Comprehensive Management Platform

Simply Stakeholders offers an end-to-end solution for mapping, communicating with, and tracking stakeholder relationships. Its strength lies in centralizing all stakeholder information and interactions, creating an organizational memory that transcends individual product managers.

The platform's stakeholder mapping functionality automatically generates influence/interest grids based on interaction patterns and formal roles. Its communication tracking ensures no stakeholder falls through the cracks during critical project phases.

Learn more at: **SimplyStakeholders.com**

Jambo: Engagement-Focused Stakeholder Platform

Jambo specializes in stakeholder engagement planning and execution. Its standout feature is the ability to create tailored engagement strategies for different stakeholder segments with automated follow-up workflows.

For products with complex stakeholder ecosystems, Jambo's analytics provide insight into engagement effectiveness, helping refine your approach over time.

Learn more at: **Jambo.cloud**

Consultation Manager: Feedback and Regulatory Compliance

For products in highly regulated industries, Consultation Manager focuses on stakeholder consultation processes that satisfy compliance requirements while generating actionable product insights.

The platform's strength is its structured consultation methods, which ensure all regulatory boxes are checked while still producing useful feedback for product development. Its audit trails provide valuable documentation when regulatory reviews occur.

Learn more at: **ConsultationManager.com**

Borealis: Enterprise-Grade Stakeholder Management

For enterprise product leaders managing complex multi-product portfolios, Borealis offers robust stakeholder management integrated with broader Environmental, Social, and Governance (ESG) considerations.

The platform's strength is connecting stakeholder management to business risks and opportunities, helping prioritize stakeholder concerns in the context of overall business strategy. Its scenario planning tools are particularly valuable for products with significant regulatory or community impact.

Learn more at: **boreal-is.com**

Conclusion

Stakeholder management is ultimately a balancing act: between different stakeholders' needs, between stakeholder input and product vision, between transparency and discretion, between collaboration and leadership.

There's no perfect formula because the human element introduces endless variability. The best product leaders develop an adaptive approach, shifting their stakeholder management style to match the specific people, products, and circumstances they encounter.

Remember that stakeholder management isn't something you do once and check off your list. It's an ongoing discipline that requires constant attention and refinement. The relationships you build and the communication patterns you establish will either become your product's greatest accelerator or its most persistent friction.

In my experience, the differentiating factor between good and great product leaders often isn't their technical knowledge or strategic brilliance; it's their ability to navigate the complex human systems that surround their products. Master this discipline, and you'll find paths to product success that remain invisible to those who focus solely on features and technology.

Now, if you'll please excuse me. I have about 47 stakeholder emails to answer before my next meeting marathon begins.

Concepts in Action: CloudTwenty

Nick Brody, the senior product manager at CloudTwenty, was tasked with launching the company's new enterprise data security platform. With just eight weeks until the planned release date, he knew stakeholder management would make or break the launch.

Defining Stakeholders

Nick's first step was creating a comprehensive stakeholder map. Beyond the obvious players (Engineering, Sales, Marketing, and the executive team), he identified several "invisible stakeholders" who would be critical to success:

- The IT security admins who would need to implement the platform
- Customer support specialists who would handle post-launch questions
- Compliance officers who could veto the product on regulatory grounds
- The assistants to the C-suite executives, who controlled access to key decision-makers

Using a power-interest grid, Nick categorized each stakeholder. The CTO and Head of Sales were clearly high-power, high-interest players who needed deep engagement. The marketing team for CloudTwenty's consumer products fell into the low-power, low-interest quadrant and simply needed occasional updates.

Building Relationships

Rather than scheduling fifteen separate stakeholder meetings, Nick organized three cross-functional sessions, forcing important trade-off discussions to happen with all perspectives represented. This not only saved countless hours but surfaced conflicts early when they could still be resolved.

Nick also distributed stakeholder relationships across his team. His technical product manager owned the relationship with the engineering team, while his UX designer became the primary contact for the customer support team. This approach scaled their stakeholder management capacity and prevented Nick from becoming a bottleneck.

Communication Strategy

For ongoing communication, Nick implemented a tiered approach:

- Weekly standups with directly involved teams
- Bi-weekly summaries for adjacent stakeholders

- Monthly executive briefings
- A dedicated Slack channel for all stakeholders to track progress

Before a major feature announcement, Nick discovered that the compliance team needed an additional two weeks to complete their review. Rather than hiding this information, he transparently shared the delay with all stakeholders, explaining the reasons and the adjusted timeline. This temporary pain preserved his long-term credibility.

Managing Expectations

When the sales director requested a complex customization feature two weeks before launch, Nick didn't simply refuse. Instead, he acknowledged the underlying need (serving a key customer segment) and proposed using configuration rather than customization. This "yes to the problem, no to the specific solution" approach maintained the relationship while protecting product integrity.

Launch Success

CloudTwenty's security platform launched with strong adoption, not because it was perfect, but because stakeholders across the organization felt ownership in its success. Nick's approach to stakeholder management had turned potential blockers into champions.

As he prepared for iterations of the platform and his next product initiative, Nick remembered: stakeholder management isn't something you do once and check off your list. It's an ongoing discipline that requires constant attention and refinement.

Execution Essentials

Defining Stakeholders

- Map stakeholders on a power-interest grid to prioritize your limited attention.
- Look beyond org charts for "invisible stakeholders" who can make or break your product.
- For highly visible products, include executive assistants as key stakeholders.
- Identify both internal (teams, execs) and external (customers, regulators) stakeholders.

Building Stakeholder Relationships

- Distinguish between strategic long-term and tactical short-term stakeholder relationships.
- Uncover stakeholders' personal motivations, not just their department's official position.
- Distribute stakeholder relationships across your team to prevent bottlenecks.
- Favor group engagement over one-on-ones to prevent information silos.
- Give more weight to stakeholders with relevant expertise for specific decisions.

Stakeholder Communication

- Use the right medium for the right message and audience (surveys, newsletters, demos).
- Follow a tiered approach: weekly for close teams, monthly for execs, quarterly for distant stakeholders.
- Tailor messages to each stakeholder group's specific concerns and interests.
- Use a RACI grid for major initiatives to clarify roles and expectations.
- Highlight key information with visual cues (colors, headers) for busy stakeholders to skim.

Navigating Pitfalls

- Say "yes" to the problem but "no" to specific solutions that don't align with strategy.
- Don't let stakeholders design your features or dictate implementation details.

- Never function as a requirements pass-through from stakeholders to developers.

- Be transparent about mistakes rather than obscuring problems.

Managing Expectations

- Define success metrics that align with product strategy, not just stakeholder wishes.

- Early communication prevents the cardinal sin of stakeholder management's "the surprise."

- Keep stakeholders informed at a level that gives confidence without creating burden.

- Look for creative compromises that address core needs while maintaining product coherence.

- Remember stakeholder management is ongoing, not a one-time task to check off.

The Role
Chapter 6

Leading Cross-Functional Teams

Cross-functional teams bring together diverse skill sets, perspectives, and working styles under one umbrella to drive product outcomes. Unlike traditional product management teams where everyone speaks roughly the same language, cross-functional teams combine engineers, designers, marketers, data scientists, and more; each with their own priorities, metrics, and ways of working.

In my online banking days, my entire team, except for the product managers, were virtually from different units. We had scrum masters and business analysts from the Agile group, software leads, developers, and testers from the technology group; all managed by different people outside my department. If you've never seen this members-on-loan arrangement before, it's called a "matrix" and it's cross-functional by design.

Product managers often don't directly manage anyone, but they "lead" product development teams. A matrix structure helps keep team members in their respective fields, allowing the PM to focus on managing the product. This includes customer outreach, go-to-market strategies, and other key areas. It's important to keep your PMs from getting bogged down in the details of managing people. Let their department heads handle that.

In this chapter, I'll share what I've learned about successfully leading complex, cross-functional teams — not just to survive, but to thrive. Because when these teams work well, they don't just build good products; they build exceptional ones that transcend what any single discipline could achieve alone.

Objectives, Goals & Indicators

Traditional product teams typically share similar skills, terminology, and objectives. When I managed my first product team many moons ago, our disagreements were usually about prioritization or feature details, not fundamental approaches to problem-solving. We all viewed the world through a product lens.

Cross-functional teams, on the other hand, bring together people who may have entirely different mental models for approaching problems. Engineers might immediately think about technical feasibility and architecture, designers about user experience and accessibility, and marketers about positioning and customer acquisition. These different perspectives are precisely where the magic happens, but also where confusion and conflict can emerge.

The key insight I've gained over the years is that cross-functional leadership isn't about making everyone think like a product manager. It's about creating an environment where these different perspectives can harmonize toward shared objectives. I've seen cross-functional teams produce breakthrough innovations specifically because they didn't all think alike.

When Sarah joined a company to lead their patient portal redesign, she assembled a team with engineers, UX (User Experience) designers, compliance experts, and clinical representatives. In early meetings, the engineers focused on backend system limitations, designers advocated for simplified interfaces, compliance experts cited regulatory requirements, and clinicians emphasized patient outcomes. Meetings felt like everyone was speaking different languages.

Instead of forcing consensus, Sarah created journey mapping workshops where each discipline contributed their expertise to different stages of the patient experience. This approach helped team members appreciate how their specialized knowledge fit into the bigger picture. The resulting patient portal not only met technical and regulatory requirements but also increased patient engagement by 47% — something that wouldn't have happened with a single-discipline approach.

Defining Very Clear Objectives

The first time I discovered a cross-functional team without clear objectives, I found a spectacular mess. They had talented people who worked hard, but they ended up with a product that tried to be everything to everyone and succeeded at nothing. I learned that cross-functional teams need clearer objectives than traditional teams, not vaguer ones.

Clear objectives provide the much-needed common ground across disciplines. When your designer, engineer, and data scientist can all articulate the same top-level goal — even if they contribute to it differently — you've established the foundation for effective collaboration. I've found that objectives should be outcome-focused rather than output-focused, giving each discipline space to contribute their best thinking.

The process of defining these objectives should be collaborative but decisive. Involve key stakeholders from each discipline to ensure buy-in, but don't fall into the trap of creating objectives so broad that they become meaningless. As the product leader, the final call on objectives is ultimately yours to make.

If a feature requires making a digital bank deposit with a paper check, don't create an objective of "making a deposit." It should be specific to the user opening your institution's mobile app, taking an image of the paper check (front and back), providing the exact amount, then submitting. The details can be helped along by your team and stakeholders. Your job was to describe the objective.

Bob led a cross-functional team at an e-commerce platform tasked with "improving the checkout experience." After initial confusion about priorities, Bob refined this broad mandate into a specific objective: "Reduce checkout abandonment by 20% while maintaining average order value within 5% of current levels." This clarity allowed the engineering team to focus on performance improvements, designers to identify and fix UX friction points, and the analytics team to build proper measurement systems. When the team exceeded their goal by reducing abandonment by 24%, each discipline could clearly see how their contributions had led to this shared success.

Establishing Key Performance Indicators

If objectives are the destination, KPIs (Key Performance Indicators) are the map and compass. In cross-functional environments, well-designed KPIs serve multiple crucial purposes: they translate abstract objectives into measurable outcomes, they provide a shared language for progress, and they help different disciplines understand how their work contributes to the whole.

I've learned that effective KPIs for cross-functional teams should be balanced across disciplines while still being cohesive. For example, if you're building a new customer onboarding flow, you might track engineering metrics like page load speed, design metrics like task completion rates, and business metrics like conversion rate; all of which contribute to the overall objective of improving onboarding success.

The most important quality of cross-functional KPIs is that they focus on outcomes rather than outputs. Tracking the number of features shipped or design assets created misses the point; what matters is the impact those contributions have on user and business outcomes.

Miguel's cross-functional team at a streaming service was tasked with increasing subscriber retention. Rather than allowing each discipline to focus solely on their traditional metrics (like code quality or visual polish), he established three shared KPIs: 28-day retention rate, average weekly viewing hours, and content discovery engagement. This approach ensured that when engineers worked on recommendation algorithm improvements, they focused on aspects that would impact these specific metrics rather than general performance enhancements. Similarly, the content team prioritized recommendations that drove weekly viewing habits rather than just click-through rates. This shared measurement method helped the team improve retention substantially and in only two quarters.

Reviewing and Adjusting Goals

The only thing more dangerous than having no goals for a cross-functional team is having unchangeable goals. I've witnessed promising initiatives fail because leadership rigidly stuck to original objectives despite new information suggesting a different approach. Cross-functional teams exist to solve complex problems, and complex problems often reveal new dimensions as you work on them.

Regular reviews of goals and KPIs should be a scheduled part of your cross-functional rhythm. I recommend quarterly deep dives coupled with monthly check-ins. These sessions aren't just about reporting numbers; they're opportunities to question whether you're still pursuing the right objectives given what you've learned.

The adjustment process should balance stability with adaptability. Too much change creates whiplash; too little ignores reality. When adjustments are needed, make them decisively and transparently, explaining the rationale to maintain team trust and alignment.

Meetings & Communication

The most productive cross-functional meetings create space for both convergent and divergent thinking and communication. Start with divergence — allowing different disciplines to bring their unique perspectives — then guide the team toward convergence on decisions or next steps. Without this deliberate structure, meetings often become either unfocused discussion forums or premature decision-making sessions that ignore important disciplinary insights.

When introducing a new feature, everyone has their own approach to implementing it. I often let the tech team explain how they'd design the user interface (UI, usually handled by the UX team) in a way that simplifies database management. Then, I'd let the UX team share any additional fields they might want to track in the database to ensure accessibility compliance (fields in a database are typically the domain of the tech team). Finally, I'd involve the business team to discuss any compliance mandates. Through this collaborative process, we reach a consensus, generate fresh ideas, and explore new perspectives. Ultimately, we build the best version of the feature.

Establish Clear Channels of Communication

I used to believe that more communication was always better until my first large cross-functional team drowned in email messages and meetings. The revelation came when a frustrated engineer told me, "I spend so much time communicating about the work that I can't actually do the work." That's when I realized that cross-functional communication needs to be thoughtful, not just abundant.

Effective cross-functional communication starts with intentional channel selection. Each team should have clearly defined channels for different types of communication: synchronous vs. asynchronous, decision-making vs. information sharing, urgent vs. non-urgent. Document these channels and their purposes so everyone knows where to go for what.

I personally like wikis, such as Confluence, Microsoft Teams, or Notion sites, where teammates and cross-functional partners can all contribute and comment in a space considered a communication hub.

Perhaps most importantly, cross-functional teams need regular but efficient synchronization points. I've found that short, daily standups combined with weekly more in-depth discussions strike the right balance for most teams. The daily touchpoints prevent silos from forming, while the weekly discussions allow for more substantive cross-discipline problem-solving.

Bonnie inherited a cross-functional product team with communication problems. Designers felt left out of technical decisions, engineers complained about constantly changing requirements, and product managers were frustrated by slow execution. She implemented a communication method that included: a daily 15-minute standup for blocking issues, a weekly 90-minute working session for collaborative problem-solving, and a bi-weekly demo for stakeholders. She also established clear documentation practices in the team wiki for decisions and designated specific Slack channels for different types of discussions. Within a month, team members reported spending 30% less time in meetings while feeling better informed about project status.

I've been saying for years that I'm going to someday write two humorous books: "My Life in Elevators" and "Death by Meetings."

Example: MediSync Solutions

MediSync Solutions, a healthcare software provider, struggled with communication breakdowns on their patient monitoring platform team. Engineers complained about constantly changing requirements, product managers couldn't get timely updates on development progress, and the compliance team felt left out of critical decisions.

CTO Rachel Torres recognized the problem immediately. "We have talented people sending messages everywhere: email, Slack, text, hallway conversations, but nothing is organized or accessible when needed," she explained to the executive team. "Our cross-functional team is drowning in communication noise."

Rachel implemented a structured communication method with clearly defined channels for different types of information:

- She established a central documentation wiki where all project requirements, design decisions, and technical specifications lived, organized by feature. This became the single source of truth that everyone could reference.

- The team created specific Slack channels with clear purposes: #patientmon-announcements for official updates, #patient-mon-decisions for documenting choices made, #patientmon-questions for queries, and #patientmon-watercooler for team building.

- Rachel instituted a meeting cadence: 15-minute daily standups focused only on blockers, weekly 60-minute working sessions for problem-solving, and monthly review meetings for broader updates. Each meeting type had a consistent agenda template and documented outcomes.

- She implemented a "decision log" that captured what was decided, by whom, and the rationale; putting an end to revisiting settled matters due to poor documentation.

- The team developed an urgent/non-urgent communication protocol, clarifying when to use which channel based on the time sensitivity of the information.

Three months after implementing these changes, development velocity increased by 40%. A compliance specialist noted, "Before, I'd miss critical design decisions and have to force costly changes later. Now I know exactly where to look for updates and when to provide input."

Product manager Sam Armstrong observed, "The noise has disappeared. When I need engineering input, I know where to ask. When I update requirements, everyone gets notified through the right channel. We're finally working as one team instead of parallel silos."

Rachel's key insight was that cross-functional communication needs to be intentional, not abundant. By establishing clear channels with specific purposes, the team eliminated the chaos of scattered communication and created an environment where different disciplines could effectively collaborate.

Encourage Open, Unabashed Dialog

Clear communication channels are necessary but insufficient. They must be filled with honest, constructive dialog. Some of the best product insights I've ever encountered came from quiet team members who initially hesitated to speak up in cross-functional settings. Creating an environment where everyone feels empowered to contribute is essential, not optional.

Open dialog in cross-functional teams requires addressing power dynamics and communication style differences. Engineers might be direct and solution-focused, while designers might communicate more visually or narratively. As the leader, you need to validate and translate between these styles, ensuring each discipline feels heard and understood.

I've found that structured dialog techniques can be particularly effective. Methods like round-robin input gathering and anonymous idea submission can bring out perspectives that might otherwise remain hidden.

One of my all-time rock stars was a brilliant engineer named Priyanka. She was usually very quiet in meetings, but one day, while working on a super complex design, I wanted to hear from everyone. I wanted every possible perspective I could get. There was a lot at stake, possibly even some jobs.

I finally called on her, and I could tell it was really tough for her to speak up. After a few seconds (and me giving her a little nudge), she started sharing some really insightful ideas that the whole team had never thought of. Once she felt like we wanted more, she became even braver and shared some of the best ideas of that meeting. Now, she's a regular leader in many meetings and is super valuable to the organization.

Do you want to find a rock star? Look for those folks in the darkness, those silent during meetings, and empower them. Look down the list of attendees and be sure to ask those who have not yet spoken at the meeting's end, "Tom, do you have anything to add? Your opinion matters as well." You will gain so much respect and, slowly, they will speak up and may even become tomorrow's leaders. I've seen it countless times.

Are You Bringing Context?

If there's one communication skill that separates good cross-functional leaders from great ones, it's the ability to facilitate effective presentations across disciplines. Your meetings serve as critical knowledge transfer points between specialties that might otherwise struggle to understand each other's work.

Effective cross-functional knowledge sharing focuses on why and what before how. Start with the business or user need, move to the proposed solution, and only then dive into implementation details. This structure ensures everyone understands the context before getting lost in discipline-specific execution plans.

DO NOT jump into what a feature is supposed to do until you have given your team a full context of "why." Without the why, they won't give a damn because you didn't give them the courtesy of context. Give them the credit for helping you build the best solution. That's what they're there to do.

I encourage teams to use a standard slide deck template that includes sections for background, objectives, proposed approach, expected outcomes, and open questions. This consistency makes it easier for specialists to understand items outside their domain and facilitates better cross-functional discussion.

Here's an example, using a three-part meeting approach:

Part 1: Context and Purpose

- What problem are we solving?
- Why does it matter to our business and customers?

- What's at stake if we do nothing?

Part 2: The Approach

- What options should we consider?
- Why should we choose a specific direction?
- What trade-offs would we be making?

Part 3: Next Steps and Decisions

- What specific actions do we need from each other?
- What decisions need to be made today?
- What does success look like?

This structure works because it mirrors how our brains process information. We need context before details, and implications after facts. Think of it like preparing a meal. You set the table before serving food and you don't start dessert before everyone's finished their main course.

Make Numbers Tell Stories

Raw data is like uncooked ingredients; nutritious but hard to digest. When presenting metrics, transform them into visual stories that illuminate rather than overwhelm. Layoff the Excel worksheets! Nobody wants to see it.

Remember the golden rule: Every chart should answer exactly one question. If you're showing a trend line of user adoption alongside feature usage segmented by user type with an overlay of support tickets, you've created a visual traffic jam where insights get lost in collisions.

Instead, think like a good journalist: What's the headline? What's the supporting evidence? Create visual hierarchy that guides eyes to what matters most. Your complex spreadsheet might be a masterpiece of analysis, but in a meeting, it's about the insight, not the calculation.

Speak Human, Not Jargon

Every department has its own dialect of business-speak, but cross-functional meetings require a common language. Industry acronyms and technical terms are like inside jokes. They create belonging for those in the know and alienation for everyone else.

This doesn't mean dumbing things down. Rather, it means being respectful enough to translate concepts into terms everyone understands. If you must use specialized language, provide brief definitions. Think of it like visiting a foreign country. A few translated phrases go a long way toward building good relationships.

When I'm new to a product or line of business, I tell everyone, "Speak to me like I'm a third-grader! I catch on quickly, but not if you treat me like I've been on your team for 10 years. I appreciate it, but don't."

Make Decisions, Not Presentations

The true measure of a good presentation isn't applause; it's alignment and action. Design your slides as decision-making tools, not information dumps.

Meetings that end with clear decisions and assigned actions aren't just more productive; they're less likely to need follow-up meetings. Like a well-designed product, a well-designed presentation anticipates user needs and facilitates desired outcomes.

Suggestion: End each key part of your meetings with what military strategists call a "decision-forcing function" — a moment that requires active choice rather than passive consumption. This might look like:

"Based on these user metrics, we have three options for the checkout flow redesign. We need to decide **today** which direction to pursue so the design team can start prototyping tomorrow."

This approach is the presentation equivalent of "Call to Action" buttons in digital marketing. Don't just inform; prompt specific action.

However, some decisions can't be made immediately due to missing information or stakeholders. Rather than letting these derail your meeting, create a visible "parking lot" for issues that need to be addressed later.

This works like a pressure release valve, acknowledging important questions while maintaining forward momentum on decisions that can be made today. It's similar to how smart product teams handle feature requests; recognize them, document them, but don't let them hijack the current work in progress.

Foster a Collaboration Culture

Culture might seem like a soft concept compared to objectives and KPIs, but I've found it's the determining factor in cross-functional success. A team with perfect processes but a poor collaboration culture will inevitably underperform compared to a team with an authentic commitment to working across boundaries.

I mentioned the importance of open and honest communication earlier. This is different from talking to individuals. It's about not showing favoritism to certain teams, like the UX team, just because you're friends or you know more about user experience. It's about making sure every team feels heard and valued.

Building a collaborative cross-functional culture starts with modeling the behavior yourself. Demonstrate curiosity about other disciplines, publicly acknowledge the value of diverse perspectives, and visibly incorporate input from different specialties into decisions. Leaders who show genuine respect for all disciplines set the tone for the entire team.

Organizational structures and incentives must support collaboration, not just pay lip service to it. Ensure that performance reviews and recognition systems reward cross-functional contributions, not just excellence within a discipline. I've seen promising initiatives fail because team members were ultimately evaluated only on discipline-specific metrics, despite rhetoric about collaboration.

Wei joined a media company to lead a cross-functional team building a new content platform, but found the existing culture strongly siloed. Engineers, designers, and content creators barely interacted outside formal meetings. Wei introduced cross-discipline shadowing, where team members spent half a day each month observing someone from a different specialty. Engineers sat with content creators to understand their workflow challenges, designers shadowed data analysts to better understand user behavior patterns, and product managers spent time with QA specialists to appreciate testing complexities. This simple practice built empathy and sparked numerous improvements as team members gained firsthand experience of different perspectives.

Keeping Connections

Personal connection is the invisible thread that holds cross-functional teams together, especially when physical proximity isn't possible. I've found that these connections don't have to be elaborate or time-consuming, but they do need to be intentional. Simple practices like starting meetings with brief check-ins or creating virtual spaces for casual interaction can maintain the human bonds that facilitate cross-discipline collaboration.

I have a little ritual for every meeting I host. It starts with a few minutes of casual, get-to-know-you chat (while waiting for everyone to show up). I'm a bit of a goofball and love to make people laugh. It's a great way to break the ice and keep everyone engaged. In fact, I've heard people say that my meetings are the only ones they look forward to all day. Sometimes, they even come to work just to be in one of them!

The most effective cross-functional teams balance formal working relationships with authentic personal connections. This doesn't mean everyone needs to be best friends, but it does require seeing team members as whole people, not just functional representatives of their discipline. I make a point of knowing something about each team member's interests outside work and creating opportunities for people to connect beyond their professional identities.

There was this attorney from Legal on one of my teams. She was super strict and by-the-book, and it was sometimes tough to work with her. But guess what? One day, I told her about my plantar fasciitis, and to my surprise, she had it too! She was such a sweetheart and super nice, and we became really good friends and even better work partners. I think everyone was surprised, but I didn't care. I saw past her tough exterior and saw the real person.

Promote Psychological Safety

Amy Edmondson's research on psychological safety revolutionized how I approach cross-functional leadership. I realized that for specialists to contribute effectively outside their comfort zones, they need to feel safe taking risks, asking questions, and occasionally being wrong.

Psychological safety is particularly important in cross-functional teams because team members regularly operate at the edges of their expertise. The designer who hesitates to ask a "dumb question" about the database architecture might miss a critical insight that could improve the user experience. The engineer who won't admit confusion about user research findings might build technically perfect features that solve the wrong problems.

Building psychological safety starts with normalizing vulnerability. As the leader, openly acknowledge what you don't know, ask questions that might seem basic, and thank people who raise concerns or point out potential issues. Create explicit norms that celebrate constructive disagreement and thoughtful questions.

Product leadership demands patience: the wisdom to nurture others' growth, the generosity to elevate their achievements, and the courage to stand firmly beside those who haven't yet weathered the storms you've survived. True leaders don't simply direct from experience; they create safe harbors where others can gain their own.

Celebrate Successes

I've found that celebration isn't just a nice-to-have morale booster. It's an essential part of the cross-functional flywheel. Thoughtful celebration reinforces shared purpose, acknowledges diverse contributions, and builds the relationships that fuel future collaboration.

Effective celebration in cross-functional contexts requires highlighting both individual discipline excellence and collective achievement. I make a point of specifically acknowledging how different specialties contributed to outcomes: "The stellar interface Sophia designed, powered by Mohamed's incredible database redesign, made this breakthrough possible. You guys are the bestest!"

Beyond formal recognition, create space for teams to share their pride in their work. Demo days, showcase events, and even simple chat channels dedicated to sharing wins all provide opportunities for team members to see the impact of their collaborative efforts.

Manny led a cross-functional team that successfully launched a complex IoT platform after 18 months of development. Rather than a generic celebration, he organized a "journey showcase" where each team presented their biggest challenge and how they overcame it. The engineering team demonstrated how they solved a particularly tricky connectivity problem, the design team showed the evolution of the interface from early sketches to final product, and the product team shared surprising user insights that had redirected the project mid-course. This approach not only celebrated success but reinforced how the different disciplines had contributed to the outcome. The team carried this integrated perspective into their next project, delivering it much faster than the original estimate.

Conclusion

Leading cross-functional teams certainly isn't the easiest thing in product leadership. The complexities of diverse perspectives, communication styles, and priorities create challenges that single-discipline teams simply don't face. There have been days when I've enviously eyed colleagues leading more homogeneous teams with their seemingly straightforward alignment.

But I've never regretted choosing this path. The products created by well-led cross-functional teams don't just incrementally improve on what came before. They often represent step-changes in how problems are solved. The diversity of cross-functional teams, when properly channeled, is a superpower.

Remember that leading cross-functional teams is ultimately about orchestration rather than control. Your job isn't to be the smartest person in every discipline represented on your team; it's to create the conditions where each specialist can contribute their best work toward shared objectives. When you get this right, the results speak for themselves.

I am alive most when I am in a cross-functional meeting. You might think it's because, being an engineer, I can relate to that side of the team. Okay, sure. Truly it's because I have been on all sides, doing most every job, and when the product sings, so do we all.

Concepts in Action: CargoTrac Technologies

CargoTrac Technologies, a mid-sized software company, was struggling with its latest product launch. Their new cargo tracking application was six months behind schedule, over budget, and missing key features. The engineering team blamed unclear requirements, designers complained about technical constraints, and marketing felt disconnected from development.

CEO Evonne Leigh knew something had to change. The traditional department-led approach wasn't working for complex products that required expertise from multiple disciplines. After reviewing several management books, she decided to reorganize the company around cross-functional teams and brought in Alex Madsen, an experienced product leader, to guide the transformation.

Clear Objectives and KPIs

Alex's first action was to gather team leaders from engineering, design, marketing, and customer support to define clear objectives for their troubled cargo tracking application. Instead of the vague goal of "improving shipment visibility," they established specific objectives:

"Reduce time to locate shipments by 40% while increasing tracking accuracy by 25% and supporting all transportation modes."

This clarity gave everyone a shared vision. The engineering team could focus on performance improvements, designers on usability, and marketers on messaging that highlighted these benefits.

Alex then established balanced KPIs that crossed disciplines:

- Cargo location time (design and engineering)
- Tracking accuracy rates (engineering and customer support)
- User satisfaction scores (design and marketing)
- Feature adoption rates (marketing and engineering)

These metrics kept everyone accountable to outcomes, not just their disciplinary outputs.

Effective Communication Channels

To address ongoing communication issues, Alex established a clear communication method:

- A daily 15-minute standup for critical updates

- Weekly 90-minute working sessions for collaborative problem-solving
- Monthly reviews of objectives and KPIs
- A team wiki where all decisions and documentation lived

The team created designated Slack channels for different conversation types: #cargo-decisions for formal decisions, #cargo-ideas for brainstorming, and #cargo-social for team building.

Fostering Open Dialog

During an early design review, Alex noticed that one of the engineers, Thomas, seemed hesitant to speak up despite clearly having reservations about a proposed feature. Alex paused the meeting and said, "Thomas, you seem to have thoughts on this approach. I'd really value hearing your perspective."

Thomas revealed that the proposed tracking interface would create significant database complexities that would slow performance. Rather than forcing a quick resolution, Alex had the team spend time understanding both the design needs and technical constraints. This collaborative approach led to a solution that balanced user experience with technical feasibility.

Alex established a meeting structure that created safe spaces for different communication styles:

- Individual reflection time before key discussions
- Round-robin input gathering to ensure all voices were heard
- Anonymous idea submission for contentious topics
- Clearly designated decision points

Building Context

Before each major feature discussion, Alex insisted on starting with context. During one pivotal meeting about the mobile tracking app, he used a three-part approach:

Part 1: Context and Purpose

"Our logistics managers spend 70% of their time in warehouses and transit hubs but only 20% of them use our current mobile app. This costs us approximately $2 million in lost productivity annually."

Part 2: The Approach

"We have three options: optimize the current app, build a new simplified app focused on scanning and location tasks, or create a progressive web app that works across devices."

Part 3: Next Steps and Decisions

"Today we need to select an approach so the design team can start prototyping. Success means increasing mobile usage by at least 40% within three months of launch."

This structure ensured everyone understood why the feature mattered before diving into implementation details.

Creating Psychological Safety

When a major feature failed in testing, Alex surprised the team with his response. Instead of looking for someone to blame, he opened the retrospective by sharing his own mistake: "I rushed us to pick a technical approach without giving proper time to explore alternatives. What can we learn from this?"

This vulnerability created space for honest assessment. The team identified several issues across disciplines that had contributed to the failure and developed a plan to address them. The feature was successfully redesigned and became one of the platform's most popular capabilities.

Celebrating Success

Three months after the transformation began, the cargo tracking application launched with overwhelmingly positive customer feedback. Rather than a generic celebration, Alex organized a "journey showcase" where each discipline highlighted their contributions and challenges overcome:

- The engineering team demonstrated how they reduced database calls by 60% while improving reliability.
- The design team showed the evolution of the interface from early sketches to final product.
- The customer support team shared how their input had shaped the notification system, reducing expected support tickets by 30%.

This approach not only celebrated success but reinforced how different disciplines had contributed to the shared outcome.

Results

One year after Alex implemented the cross-functional approach, CargoTrac Technologies had transformed:

- The cargo tracking application became the company's fastest-growing product, with customer satisfaction scores 40% higher than industry average

- Development cycles decreased from 8 months to 10 weeks
- Employee satisfaction scores increased across all departments
- Cross-functional team members reported greater career satisfaction and learning

Evonne reflected on the transformation: "We used to think success came from having the best individual experts. Now we understand that our greatest strength lies in how effectively these experts collaborate across disciplines. The whole truly is greater than the sum of its parts."

Execution Essentials

Setting Clear Objectives

- Define specific, measurable objectives that provide common ground across disciplines, focusing on outcomes rather than outputs.
- Create balanced KPIs that cross disciplines, ensuring each team member sees how their work contributes to shared goals.
- Schedule regular reviews (quarterly deep dives with monthly check-ins) to adapt objectives based on new information.
- Make objectives outcome-focused to give each discipline space to contribute their best thinking.

Effective Meeting Management

- Structure meetings with both divergent thinking (allowing different perspectives) and convergent thinking (guiding toward decisions).
- Use a three-part meeting approach: Context/Purpose, Approach, and Next Steps/Decisions.
- End meetings with clear decisions and assigned actions to reduce follow-up meetings.
- Create a visible "parking lot" for issues that need to be addressed later to maintain forward momentum.
- Keep presentations focused on the "why" and "what" before diving into the "how" of implementation.

Communication Channels

- Establish clearly defined channels for different types of communication (synchronous vs. asynchronous, decision-making vs. information sharing).
- Consider using wikis or other collaborative platforms where all team members can contribute and comment.
- Schedule regular but efficient synchronization points (daily standups, weekly deeper discussions).
- Document communication protocols and their purposes so everyone knows where to go for what.
- Create specific channels for different conversation types (decisions, ideas, social).

Encouraging Open Dialog

- Address power dynamics and communication style differences across disciplines.
- Use structured dialog techniques like round-robin input gathering to ensure all voices are heard.
- Look for quiet contributors and actively invite their participation.
- Create psychological safety by modeling vulnerability as a leader.
- Establish norms that celebrate constructive disagreement and thoughtful questions.

Presenting Information Effectively

- Transform raw data into visual stories that illuminate rather than overwhelm.
- Ensure every chart answers exactly one question with clear visual hierarchy.
- Avoid discipline-specific jargon or provide brief definitions when necessary.
- Design presentations as decision-making tools rather than information dumps.
- Use "decision-forcing functions" at key points to prompt active choices.

Building a Collaborative Culture

- Model curiosity about other disciplines and publicly acknowledge diverse perspectives.
- Ensure performance reviews and recognition systems reward cross-functional contributions.
- Consider cross-discipline shadowing to build empathy and understanding.
- Balance formal working relationships with authentic personal connections.
- Create intentional spaces for team members to connect beyond their professional identities.

Celebration and Recognition

- Highlight both individual discipline excellence and collective achievement in celebrations.
- Create opportunities for teams to share their pride in their work through demos and showcases.
- Organize "journey showcases" where each team presents their biggest challenges and solutions.

- Acknowledge specific contributions from different specialties to reinforce their value.

- Use celebrations to reinforce shared purpose and build relationships that fuel future collaboration.

The Role

Chapter 7

On-site, Hybrid &
Remote Teams

This chapter isn't about painting an unrealistically rosy picture of distributed product teams. There were plenty of communication mishaps, moments of isolation, and productivity plateaus along the way. But what I discovered — and what I hope to share with you — is that with the right mindset, systems, and leadership approach, a flexible working model can actually strengthen your product organization rather than diminish it.

The key lies not in trying to replicate the old ways of working through digital means, but in embracing a new paradigm that plays to the strengths of a more fluid, adaptable team structure. It requires maturity, discipline, and intentionality from everyone involved — especially you as the leader — but the rewards can be substantial.

Promote Flexibility & Maturity

Executives demanding a return-to-office mandate for all employees won't solve anything, and employees staging a coup when they can't work in their pajamas isn't helpful either. The workforce needs a realistic approach to the new way business operates.

The One-Size-Fits-Nobody Reality

Soon after COVID and early in our transition to a hybrid model, I made what I now recognize as a classic blunder: I assumed everyone on my team was just like me. As someone who thrives on in-person collaboration and the energy of the office environment, I initially structured our hybrid approach to maximize collaborative on-site days. The pushback was swift and illuminating.

"I get more done in three hours at home than in a full day at the office," explained one of our senior PMs who managed our platform team. "The constant interruptions and background noise make it impossible for me to focus on strategic thinking." Meanwhile, our mobile initiatives leader, countered: "I miss the spontaneous conversations. It's so much easier to align with design and engineering when we can gather around the same whiteboard."

What became abundantly clear was that preferences around work environments aren't just personal quirks. They're often deeply tied to working styles, life circumstances, and even personality types. Some team members are genuinely more productive in solitude, while others draw energy and inspiration from social interactions. Some have home situations conducive to deep work, while others struggle with distractions outside the office.

Emotional maturity in leadership means recognizing and respecting these differences rather than imposing a one-size-fits-all solution. It means creating systems flexible enough to accommodate various working styles while maintaining team cohesion and productivity goals.

Take the case of one VP of Product, who insisted everyone return to five days in the office, citing "collaboration benefits" despite clear evidence that their data science team produced significantly better work when allowed focused time at home. The result? They lost three senior team members in two months, all citing the rigid policy as their primary reason for leaving. A more emotionally mature approach would have been to analyze which types of work benefited from co-location versus which thrived in quiet, uninterrupted environments.

Inclusion Isn't Optional

"I feel like a second-class citizen when I'm dialing in and everyone else is in the room," a PM told me during a one-on-one. She had relocated to be closer to family, working fully remote while most of the team was hybrid. Her comment was a wake-up call.

Remote inclusion isn't just about making sure everyone can hear each other on calls. It's about fundamentally rethinking how we conduct meetings, make decisions, and build team culture in ways that don't inadvertently privilege those who happen to be physically present.

True inclusion for remote team members requires intentional design. Well managed teams have implemented several practices that make a measurable difference. First, they adopt a "remote-first" meeting protocol. If one person was remote, everyone logged into the video call from their own device, even if some were sitting in the same room. This eliminates the power imbalance of having remote folks peering into a conference room where they couldn't read body language or jump into the conversation easily.

Second, they establish a digital-first documentation culture. All important discussions, decisions, and context are captured in shared digital spaces, not just verbally communicated or written on physical whiteboards that remote team members cannot see. This simple shift dramatically reduces the information asymmetry that often disadvantages remote workers.

Third, they become vigilant about "presence privilege"; the subtle ways in-office employees might receive preferential treatment simply because they were physically visible to leadership. Good leaders make a point of recognizing contributions based on impact, not visibility, and ensure that career advancement opportunities are equally accessible regardless of working arrangement.

The results will speak for themselves. Remote team members will feel "fully included" and the remote retention rate will equal in-office retention.

Balancing Flexibility with Connection

While flexibility is essential, I've found that having designated on-site days creates valuable synchronization points for my teams. After experimenting with various approaches, I've settled on 2-3 mandatory in-office days per week (Tuesdays, Wednesdays, and/or Thursdays) for all local team members, with fully remote employees flying in once per quarter for extended co-working periods. Mondays just raise stress levels on Sunday evenings and is anyone really, fully in their minds on Fridays?!

These anchor points provide structure around which you're able to organize your highest-value collaborative activities: strategic planning sessions, cross-functional workshops, and relationship-building events. The predictability allows team members to plan their lives while still ensuring regular face-to-face connection.

Mark, who joined a team during the pandemic and had never worked alongside his colleagues in person, described his first in-office week as revelatory: "I finally understood why certain decisions were made the way they were. There's a whole layer of context you pick up from being physically present that's almost impossible to communicate remotely."

This balance between flexibility and structure has proved to be the sweet spot for most organizations. Teams that skewed too far toward unstructured flexibility often reported feelings of disconnection and misalignment, while those that were too rigid lost the benefits of focused deep work and personal autonomy (as well as losing key employees).

For product teams specifically, this balance is particularly crucial. The nature of product management requires both deep, uninterrupted thinking time and high-bandwidth collaboration with stakeholders. Finding the right rhythm between these modes can unlock unprecedented productivity.

Maintaining Strong Communication

When a team works across different locations and time zones, communication becomes both more crucial and more challenging. The casual conversations by the coffee machine, the quick desk drop-by, and the ability to read a room during meetings all disappear in distributed settings. What replaces them isn't just digital tools, but an entirely new approach to information sharing that requires intentionality, clarity, and technological savvy. In my time of leading distributed product teams, I've found that communication excellence is the single greatest predictor of team success — more important than individual talent, process maturity, or even product-market fit.

The Protocol Paradox

"But I sent you a Slack message about that three days ago!"

"I put all the details in the Jira issue's comments."

"Didn't you see my email update?"

These frustrating exchanges became all too common in my early days of distributed work. Without the natural information flow of an office environment, I was once drowning in a sea of disjointed communication channels, with critical information scattered across platforms with no clear hierarchy or protocol.

The solution wasn't adding more tools or sending more messages. It was establishing clear communication protocols that everyone understood and followed. We created a simple method that mapped types of communication to specific channels:

- For immediate needs requiring quick responses (under 2 hours): Slack or Teams direct messages

- For team-wide updates and FYIs: Dedicated Slack or Teams channels

- For decisions that needed documentation: Confluence pages with Slack or Teams notifications

- For task-specific details: Jira issues (tickets), tagging other with @mentions for attention

- For complex discussions requiring nuance: Zoom or Teams calls (scheduled or impromptu)

- For major announcements or culture building: Team-wide meetings

We also established response time expectations for each channel, with the understanding that immediate response was only expected for urgent matters. This freed team members from the tyranny of continuous partial attention — checking every platform constantly out of fear of missing something important. Hell, people these days do enough of that with their phones!

The key insight was that good communication protocols should reduce cognitive overhead, not increase it. They should clarify when, how, and where information should flow, eliminating the anxiety and inefficiency of multi-platform monitoring.

Within weeks of implementing this simple method, leaders should see a huge reduction in "dropped ball" incidents and a significant decrease in reported communication stress. Team members will no longer have to guess where important information might be hiding.

One last thing on this topic... Do not allow team members (including cross-functional team members) to comment on Jira stories inside subtasks or epics. We have a saying, "That's where comments go to die and never be seen again." Instead, comments involving a story (ticket or issue) should be on the main story, not buried layers deep for nobody to see.

Technical Proficiency Isn't Optional

"I can't get my microphone to work" or "How do I share my screen again?" quickly became phrases that made my heart sink in the early days of remote collaboration. Technical difficulties might seem like minor inconveniences, but they compound dramatically at scale, wasting precious synchronous time and eroding team patience.

I realized that technical proficiency with remote collaboration tools wasn't a nice-to-have. It was a fundamental job skill for the modern product manager. I gave mandatory training sessions for the entire team on our core tools: Outlook, Skype (at the time), Teams (our new platform), and our project management software (Jira). These weren't cursory overviews but deep dives into power-user features and troubleshooting.

The investment paid dividends immediately. Meetings started on time, collaboration became smoother, and the team discovered features that actively improved our workflow. One of our more technically challenged PMs actually became the team's Teams virtuoso after learning some advanced features, transforming our virtual whiteboarding sessions from clunky imitations of in-person work to something more powerful and flexible.

Beyond the tools themselves, we cultivated meta-skills for remote communication: speaking concisely, active listening without visual cues, explicit turn-taking, and checking for understanding without the benefit of full body language. These soft skills proved just as important as technical prowess.

An important thing that I did was demand that anyone attending any of my teams' meetings must have a headset with a microphone (boom) that extended in front of their mouths; like a Polycom or Jabra with noise-cancelling technology. I did not (and still don't) allow anyone to attend my meetings by speaking over their computer's microphone and listening with their computers's speaker(s) — in other words, their computer's speakerphone. They ultimately talk over others, you hear them banging keys on their keyboards, they don't let others get word in edgewise, and they disrupt the flow of the meeting. It's insanity, immature, and very annoying.

The Meeting Minimalist

If there's one universal lesson from the shift to distributed work, it's this: we were having too many meetings all along. The transition simply made the cost more obvious.

In our office environment, a spontaneous 30-minute meeting seemed harmless. In a distributed context, each meeting began to reveal its true cost: context switching, preparation time, and the interruption of deep work. We became ruthless meeting minimalists out of necessity, and discovered our productivity improved as a result.

I established several principles that dramatically improved my teams' meeting hygiene:

- No meeting without an agenda shared at least 4 hours in advance
- Default to 25 or 50 minutes instead of 30 or 60 to create transition time
- "Camera optional" for meetings that didn't require visual interaction
- Standing meetings required quarterly justification to continue existing
- Async alternatives (document comments, recorded updates) offered whenever possible, so people didn't have to attend when swamped
- Absolutely no speaker phones; headphones only

One of our most successful innovations was "async first, sync second"; starting with asynchronous document collaboration (back-and-forth emails for refinement) before any synchronous meeting. Team members would add their perspectives to a shared document, allowing everyone to absorb and reflect on different viewpoints before coming together to resolve differences or make decisions. This approach yielded more thoughtful input, especially from introverts who might not speak up in real-time settings, and made our synchronous time significantly more productive.

Trello, a company I admire for their remote work practices, takes this even further with "async weeks." These are entire weeks without internal meetings, dedicated to deep work and document-based collaboration. While that approach might be too extreme for many organizations, the principle of creating significant meeting-free blocks is sound.

The Professional Remote Presence

I once publicly chastised a team member for having a messy bedroom background during an important stakeholder call. It was a knee-jerk reaction that I'm not proud of, but it highlighted an important truth: visual context affects perception, fairly or not.

When we interact virtually, our environments become part of our professional presence in ways that didn't matter in shared office spaces. The cat walking across the keyboard, the unmade bed in the background, or the poor lighting that makes us look like we're joining in from a submarine; these elements affect how our communication is received.

We established baseline expectations for video presence during external-facing meetings: neutral backgrounds (real ones, not fake penthouse apartments), adequate lighting, stable internet connections, and appropriate attire. The standard wasn't about enforcing formality but ensuring that the medium didn't detract from the message.

For internal team communications, we adopted a more relaxed approach, acknowledging that seeing authentic glimpses of each other's lives could actually strengthen team bonds. Carlos's toddler making a surprise appearance or Dianne's enthusiastic dog became welcome moments of humanity rather than professional faux pas. But please, under absolutely no circumstance can there be dogs barking, children screaming, loud televisions, screeching train wheels, honking cars and sirens, or blaring music.

The guideline became: "External calls are like bringing clients to the office; internal calls are like team members visiting your desk." This simple method helped people navigate the appropriate level of formality without rigid rules.

What surprised me was how these standards elevated the quality of our virtual interactions. When basic presentation issues were handled, people could focus entirely on content and connection rather than being distracted by environmental factors.

Example: SensiPoint

SensiPoint, an IoT software company based in Los Angeles, transformed their distributed work culture by mastering the communication challenges mentioned earlier.

When the pandemic hit, CEO Matt Vargus initially struggled with the communication chaos. Team members were sending crucial updates across Slack, email, and their project management tool with no rhyme or reason. After several important product decisions fell through the cracks, Matt implemented what he called a "channel protocol"; a simple system mapping specific types of communication to the right platforms. Immediate needs went to direct messages, team updates to dedicated channels, and all decisions were documented properly in Jira. Response time expectations were clearly established for each channel.

The company also invested heavily in technical training. Jason Park, their senior product manager, went from being the guy who always had his microphone on mute to becoming their virtual meeting expert. They required proper headsets with noise-canceling technology for everyone, eliminating the distraction of keyboard clicking and background noise during important discussions.

SensiPoint became ruthless about meeting efficiency. Their "async first, sync second" rule meant product teams collaborated on documents before jumping into calls, making their actual meeting time far more productive. They cut their standing meetings by 40% and required agendas (simple bullet points) for everything.

For external meetings, SensiPoint established reasonable standards for video presence — neutral backgrounds and decent lighting — while keeping internal meetings more casual and human. This balanced approach helped them maintain professionalism with clients while building genuine connections within the team.

Nine months after implementing these communication practices, SensiPoint saw measurable improvements in their IoT product delivery timelines and team satisfaction scores. Their approach to remote communication didn't just preserve their previous effectiveness — it actually enhanced it.

Requiring Discipline and Accountability

The elephant in the room with distributed work is always the same: "How do I know people are actually working?" It's a question that reveals more about the asker than the asked.

In transitioning to hybrid and remote arrangements, I've found that accountability doesn't diminish; it transforms. The visible trappings of productivity (time spent at a desk, participation in meetings) give way to more meaningful measures: outcomes delivered, problems solved, and value created. This shift demands greater discipline from team members and greater trust from leaders, creating a more mature professional relationship that, when properly nurtured, yields superior results compared to traditional supervision models.

The Self-Disciplined Professional

"Working from home isn't for everyone," is a refrain I've heard from countless executives skeptical of remote work. What they really mean is: "Not everyone has the self-discipline to stay productive without direct supervision."

There's some truth to this concern. The office environment provides external structure and accountability that disappears at home. What many leaders miss, however, is that this is an opportunity to develop and recognize self-discipline as a valuable professional skill, not a reason to force everyone back to assigned desks.

The most effective approach proves to be a combination of clear expectations and demonstrated trust. Define outcomes that need to be achieved and key performance indicators that should be tracked, then give people substantial autonomy in how they structured their work to meet those standards.

Jason, who initially struggled with the transition to working from home, developed a "mock commute" routine — walking around the block before sitting down at his desk — that helped him mentally transition into work mode. Elena established strict "do not disturb" hours for deep work, resulting in an increase in her feature specification output. These personal systems emerged organically when given the space to develop.

Far from enabling slacking off, this focus on self-discipline often revealed which team members were truly driving results versus those who had been skating by on "presenteeism" — looking busy without producing meaningful outcomes.

A product leader and friend shared a similar observation with me: "Remote work didn't create performance problems; it revealed them. The people who were quietly carrying the team became much more obvious when we could only see outputs, not activities."

The Office as a Privilege

One unexpected insight from our hybrid work model was a complete reversal in how we viewed office time. Rather than treating remote work as a special privilege granted to high performers (a common approach pre-pandemic), we began to see office access as the scarce resource and valuable perk.

Our limited designated office days became highly intentional. Team members arrived prepared, knowing exactly which conversations needed to happen face-to-face and which collaborations would benefit from co-location. The previous habit of showing up to mindlessly check email from a desk was replaced by purposeful interaction and energy.

This shift required a parallel change in expectations around professional presence. When in the office, team members needed to be truly present: engaged, prepared, and bringing their best selves. This wasn't about superficial aspects like attire (though basic grooming and appropriate dress remained expectations), but about mental presence and preparation.

"If you're coming to the office just to take Zoom calls from a different location, that's a wasted opportunity," became our guiding principle. This framing helped people see that both remote and in-office modes had specific advantages that should be leveraged intentionally.

The social and collaborative aspects of office time took on renewed importance. We prioritized relationship-building activities, creative workshops, and complex problem-solving sessions for our in-person days, while reserving focused execution work for remote days when possible.

This approach transformed our office from a daily requirement to a valuable resource that enhanced specific types of work: a shift that increased both the perceived value and the actual usefulness of our physical workspace.

The Rhythm of Accountability

While trust is fundamental, regular check-ins provide the rhythm that keeps distributed teams synchronized. The key insight that transformed my approach was distinguishing between check-ins designed for accountability and those meant for collaboration or problem-solving.

I established a simple system of asynchronous daily updates where team members answered three questions:

- What are your latest accomplishments?

- What are your focused on at the moment?
- What obstacles are in your way?

These updates weren't for micromanagement but created transparency that allowed for timely support when needed. They took less than five minutes to complete but provided crucial continuity for a distributed team.

Beyond daily updates, we maintained weekly one-on-ones between leaders and their direct reports, focusing more on progress toward goals, personal development, and addressing any emerging issues. Monthly team retrospectives rounded out our accountability rhythm, allowing for broader reflection and process improvement.

This multi-layered approach creates frequent, lightweight touchpoints that maintain momentum without excessive meeting overhead. The consistency of these check-ins — not their duration or formality — will prove to be the critical factor.

Conclusion

Looking back on our journey from a traditional co-located team to a high-functioning distributed model, I'm struck by how fundamentally this transformation changed my understanding of leadership itself. The skills that make for effective product leadership in this new paradigm aren't just adaptations of old approaches. They represent an evolution in how we think about team performance, communication, and trust.

The most successful product leaders I know have embraced this shift not as a temporary accommodation but as an opportunity to build more resilient, inclusive, and powerful teams. They've discovered, as I did, that the constraints of distributed work often lead to more intentional communication, more thoughtful processes, and more equitable participation.

This doesn't mean distributed work is without challenges. It requires more explicit structure, more intentional connection, and more disciplined communication than co-located work ever did. But when approached with maturity and intentionality, these challenges become catalysts for developing stronger organizational muscles.

As we look toward the future, the question isn't whether teams should be remote, hybrid, or on-site. It's how leaders can create systems that harness the benefits of each modality while mitigating their inherent limitations. The teams that will thrive are those that can seamlessly flow between synchronous and asynchronous collaboration, between deep focus and high-bandwidth interaction, between structured processes and creative emergence.

The product organizations that master this multi-modal approach won't just survive in the new work landscape. They'll define it, creating competitive advantages through their ability to attract diverse talent, foster inclusive collaboration, and deliver exceptional results regardless of physical proximity.

Concepts in Action: Noviki Brass

Noviki Brass, a small to medium-sized manufacturer specializing in artisanal door handles and cabinet hardware, was struggling with their return-to-office strategy. CEO Marlene Vujicic had decreed a universal five-day office mandate, convinced that "craftsmanship needs supervision."

"We've built this company on hands-on mentorship and quality control," Marlene insisted during an executive meeting. "Our designers need to see and feel the materials."

VP of Product Omar Ramirez nodded along in the meeting, but privately worried. His team had shown wildly different productivity patterns during their temporary remote work period. Eliza, his chief hardware designer, produced her most innovative sketches from her home studio, while James, the manufacturing liaison, clearly thrived on workshop floor interactions.

When Omar analyzed team output data, he discovered something fascinating: their CAD designers produced 35% more finalized designs on work-from-home days, while their prototype team generated significantly better physical samples during co-located sessions.

"We're treating our craftspeople like identical brass fixtures," Omar realized. "But our people are individuals with different creative processes and life circumstances."

The Protocol Paradox

Communication chaos reigned in Noviki's early hybrid attempts.

"Did you see my email about the new Victorian-inspired collection?" Omar asked during a status meeting.

"No, I was looking for it in the project management system," replied Diana, looking confused.

"But I sent the sketches through the messenger app yesterday!" chimed in Marcus.

After one particularly disastrous product launch where critical feedback on handle grip ergonomics got lost between various communication channels, Omar instituted clear communication protocols:

- Immediate needs (responses needed within 2 hours): Direct messages
- Team-wide updates: Dedicated channels by collection line
- Decision documentation: Digital design system with notifications

- Task details: Project tickets with attention flags
- Complex discussions: Video or in-person meetings
- Major announcements: Weekly all-hands

Each channel came with clear response time expectations. No longer did team members anxiously monitor all platforms simultaneously, fearing they'd miss something critical.

Tech Proficiency Isn't Optional

"I can't get my camera to show the prototype clearly," mumbled designer Talia through her laptop speakers during a crucial retailer review. Five minutes of awkward troubleshooting followed while buyers waited impatiently.

Omar implemented mandatory training sessions covering their core collaboration tools. He established clear standards:

- Quality headsets required for all meetings
- Proper lighting and camera setup for product demonstrations
- Basic troubleshooting skills required for all team members

"Your remote setup is now part of our professional presence," Omar explained. "When you're showing our luxury hardware to clients, your technical setup needs to match our product quality."

The Meeting Minimalist

"I counted 22 hours of meetings last week," said Paulo, looking exhausted during his one-on-one with Omar. "I barely touched my design tools."

Noviki's solution was practical but effective:

- All meetings required agendas shared at least 4 hours in advance
- 25 or 50-minute meetings instead of 30 or 60 to create transition time
- "Camera optional" for information-sharing meetings
- Standing meetings required quarterly justification

The team adopted a practice of starting with document collaboration before scheduling live discussions. This gave everyone time to absorb information and formulate thoughts, making subsequent meetings more productive and inclusive.

Workshop as a Privilege

Rather than treating remote work as a special perk, Noviki Brass flipped the script: workshop time became the valuable resource.

"If you're coming in just to answer emails from a different desk, that's a wasted opportunity," Omar told his team. "Our in-workshop days are for hands-on collaboration and prototyping."

The company designated Tuesdays and Thursdays as "collaboration days" when most local team members came to the workshop. These days were carefully structured around physical prototyping, material selection, and testing sessions. Remote team members flew in quarterly for extended co-working periods.

"I used to resent mandatory workshop days," admitted designer Jordan. "But now that they're purposeful, I actually look forward to them. We accomplish in one focused day what used to take a week of back-and-forth in messages."

The Rhythm of Accountability

To maintain momentum across distributed teams, Noviki established consistent check-in rhythms:

- Quick daily updates answering three questions: recent accomplishments, current focus, and obstacles
- Weekly one-on-ones between managers and direct reports
- Monthly team retrospectives to address process improvements

The daily updates weren't for micromanagement but created transparency that allowed for timely support. They took just minutes to complete but provided crucial continuity.

"I was skeptical of daily updates at first," confessed senior craftsman Miguel. "But they've actually reduced my stress. I don't worry about whether my detailed metalwork is visible anymore, and I get help with technical blockers much faster."

Inclusion Isn't Optional

"I feel disconnected from material decisions," remote team member Leila had complained early in the transition. "Everyone in the workshop gets to touch the brass samples while I'm just looking at images."

Noviki adopted innovative solutions: they sent material sample kits to remote team members and used high-definition cameras for detailed texture discussions. For important meetings, if one person was remote, everyone logged into the video call individually, even those in the workshop.

They also established a digital-first documentation culture. All important discussions and material decisions were captured in shared digital spaces, with detailed photography and specifications that remote team members could fully access.

"The biggest mindset shift was becoming vigilant about 'presence privilege,'" Omar explained to new hires. "We now recognize team contributions based on design impact, not physical presence."

The results were clear: remote team member retention matched that of in-office employees, and satisfaction scores were nearly identical across working arrangements.

The Self-Disciplined Professional

"Working away from the workshop isn't for hardware designers," Marlene had initially warned. But what Noviki discovered was more nuanced.

Rather than monitoring time spent at workbenches, the company shifted to outcome-based evaluation. Clear expectations were set for design quality and production deadlines, and team members were given substantial autonomy in how they structured their work.

"Remote work didn't create performance problems. It revealed them," Omar observed. "The people who were quietly carrying our design direction became much more obvious when we could only see outputs, not activities."

By embracing flexibility, establishing clear communication protocols, and creating intentional collaboration opportunities, Noviki Brass built a stronger, more innovative product development organization. Their hybrid approach wasn't about replicating old ways of working through digital means, but embracing a new paradigm that played to the strengths of both traditional craftsmanship and modern design flexibility.

Execution Essentials

Promoting Flexibility & Maturity

- Recognize that different team members have different optimal working environments based on their role, personality, and home situation.

- Analyze which types of work truly benefit from co-location versus remote settings instead of making blanket policies.

- Avoid rigid policies that treat everyone identically, as this often leads to talent loss.

- Measure productivity through outcomes and deliverables rather than hours visible at a desk.

Maintaining Strong Communication

- Establish clear communication protocols that map specific types of messages to appropriate channels.

- Define expected response times for each communication channel to reduce anxiety about missing important information.

- Implement regular but lightweight check-ins that provide transparency without micromanagement.

Technical Proficiency Standards

- Require proper audio equipment (quality headsets with boom microphones) for all team members.

- Ban computer speakerphones in meetings to prevent audio disruption and talking over others.

- Provide training on all collaboration tools, focusing on troubleshooting and advanced features.

- Establish basic standards for video backgrounds and lighting for client-facing meetings.

Effective Meeting Practices

- Require agendas for all meetings shared at least 4 hours in advance.

- Schedule 25 or 50-minute meetings instead of 30 or 60 to create transition time.

- Make standing meetings justify their existence quarterly or be eliminated.

- Designate specific days for high-value collaborative activities.

Remote Inclusion Strategies

- Adopt "remote-first" meeting protocols where everyone joins virtually if even one person is remote.

- Be vigilant about "presence privilege" by recognizing contributions based on impact, not visibility.

- Use high-quality cameras and documentation for texture and material discussions.

- Create equal career advancement opportunities regardless of working arrangement.

The Office as a Resource

- Transform office days from an obligation to a valuable opportunity for specific types of work.

- Schedule in-person days (typically Tuesday-Thursday) for highest-value collaborative activities.

- Make in-person time purposeful and focused on activities that truly benefit from co-location.

- Avoid wasting in-office time on activities that could be done just as effectively remotely.

Self-Discipline & Accountability

- Set clear outcome-based expectations rather than monitoring time or activity.

- Implement daily updates focused on accomplishments, current focus, and obstacles.

- Recognize that remote work often reveals pre-existing performance issues rather than creating them.

Professional Remote Presence

- Establish baseline expectations for video presence during external-facing meetings.

- Allow more casual and authentic interactions for internal team communication.

- Ensure remote setups reflect the quality standards of your products, especially for client demonstrations.

- Address basic presentation issues to keep focus on content rather than distracting environmental factors.

The Role

Chapter 8

Developing Future Leaders

If you've made it to a product leadership position, you've likely benefited from someone who took the time to invest in your growth. Maybe it was that VP who saw potential in you when you were just starting out, or perhaps it was a peer who pushed you to think differently about a challenging problem. Whatever form it took, that investment helped shape your career. Now that you've "made it," one of your most important responsibilities isn't driving quarterly results or shipping features; it's cultivating the next generation of product leaders.

The Approach

Before we dive deeper, it's important to understand the three approaches to boosting your team members: coaching, mentoring, and servant leadership. While they share the common goal of helping others, they operate in different ways and serve different purposes. The most effective product leaders know when to apply each approach based on the individual's needs, the situation at hand, and the desired outcome.

Let's explore each of these pillars in depth, understanding not just what they are, but how to implement them effectively in your day-to-day work as a product leader. The mastery of these three disciplines won't just elevate your team; it will transform how you approach leadership itself.

Coaching

Coaching is perhaps the most powerful — and most misunderstood — tool in a product leader's development arsenal. Unlike directing or advising, coaching creates the conditions for self-discovery. It's built on the premise that most product managers already have the answers they need within them; they just need the right questions and space to uncover those insights.

When I first attempted to "coach" team members, I was really just giving advice disguised as questions: "Have you considered prioritizing feature X instead?" This isn't coaching; it's steering with a thin veneer of inquiry. True coaching requires suspending your own agenda and trusting in the other person's capacity to find their own path.

Effective coaching begins with creating psychological safety. Your product managers need to know they can be vulnerable and explore their thinking without fear of judgment. At Facebook, Julie Zhuo (former VP of Product Design) was known for starting coaching conversations with junior PMs by sharing her own mistakes and learnings, immediately establishing that it was safe to be imperfect.

Asking the Right Questions

The hallmark of great coaching is asking questions that cause the other person to pause, reflect, and see their situation in a new light. These questions should be open-ended, non-leading, and focused on expanding thinking rather than narrowing it.

Some of the most impactful coaching questions I've used with product managers include:

- "What assumptions are you making that might not be true?"
- "If you knew you couldn't fail, what approach would you take?"
- "How would our most demanding customer view this decision?"
- "What's the perspective you're finding hardest to consider?"
- "What are three alternative ways to frame this problem?"

The key is to ask these questions and then get comfortable with silence. Fight the urge to fill the space. Some of the most profound insights come after those seemingly uncomfortable pauses when the product manager is truly wrestling with a new perspective.

In my experience, the most powerful coaching moments often come during product reviews. Rather than simply evaluating the work, use these sessions as opportunities to help PMs examine their thought processes. When one of my PMs presented a feature that didn't quite hit the mark, instead of pointing out the flaws, I asked: "Walk me through how you arrived at this solution." This simple question helped her recognize a critical step she'd skipped in understanding user needs; a realization far more valuable than any direct feedback I could have given.

Building Problem-Solving Muscles

The ultimate goal of coaching isn't just to solve the immediate problem. It's to build your product manager's problem-solving capabilities for the countless challenges they'll face in your absence. This means resisting the temptation to offer solutions, even when you see them clearly.

When a senior PM on my team was struggling with a complex prioritization decision, every fiber of my being wanted to tell him what to do. Instead, I asked him to walk me through his mental model for making these decisions. As he articulated his thinking, he identified the gaps himself. The process took longer than if I'd just given the answer, but the growth in his decision-making method was worth the investment.

Instead of telling product managers what to do in specific situations, help them understand the underlying principles that can guide their decisions across countless scenarios.

Example: TimeDex

At TimeDex, a smaller watch movement company, product leader Susan noticed her team kept bringing her fully-baked solutions instead of exploring possibilities. During one review, a PM named Tyler presented a feature that missed the mark on user needs.

Instead of critiquing the design, Susan asked, "Walk me through how you arrived at this solution." After an uncomfortable pause, Tyler realized he'd skipped crucial steps in understanding user pain points. This simple question helped him discover the gap in his process.

Later that quarter, when senior PM Jamie was struggling with a complex prioritization decision, Susan resisted giving him answers. She instead asked, "What's your mental model for making these decisions?" As Jamie talked through his thinking, he identified the weaknesses himself. Though this took longer than Susan providing the solution, Jamie developed decision-making skills he could apply to countless future scenarios.

Susan established "Thinking Thursdays" where PMs bring tough problems without solutions. The team asks questions like "What assumptions are you making that might not be true?" or "How would our most demanding customer view this?" This practice has transformed TimeDex's product development process. PMs now tackle problems more creatively and require less direct guidance from leadership.

The result? Better products built by more confident product managers who know how to navigate complex decisions independently.

Structured Feedback for Growth

While much of coaching focuses on forward-looking development, providing structured feedback on past performance is equally important. The key is delivering this feedback in a way that promotes growth rather than triggering defensiveness.

I've found the SBI (Situation-Behavior-Impact) method particularly effective:

"In yesterday's stakeholder meeting [Situation], when you presented multiple options rather than a single recommendation [Behavior], the engineering team left feeling uncertain about priorities [Impact]."

This approach keeps feedback objective and specific, making it actionable without feeling like a personal criticism. For longer-term coaching relationships, I supplement this with regular "feed-forward" sessions; focused discussions about specific behaviors the PM wants to improve in the future.

Time-Bound, Clear Outcomes

Effective coaching relationships should have defined timeframes and outcomes. Open-ended coaching without clear goals often loses momentum and impact. When I begin a coaching relationship with a product manager, we establish specific objectives and a timeline for achieving them.

For example, with a mid-level PM struggling with stakeholder management, we established a three-month coaching engagement focused specifically on this skill, with clear success metrics: successfully leading a cross-functional initiative, gathering feedback from key stakeholders, and increasing their confidence in navigating organizational politics.

Setting these parameters creates accountability for both parties and ensures the coaching relationship remains productive and purposeful rather than drifting into casual check-ins.

Mentoring

While coaching focuses on drawing out the product manager's own insights, mentoring involves more directly sharing your wisdom and experience. As a mentor, you're not just asking powerful questions; you're offering guidance based on your journey. This form of development is particularly valuable for helping product managers navigate situations they haven't encountered before.

I distinctly remember my first catastrophic product launch; a feature that solved the wrong problem and was met with user backlash. Someone I looked up to as a mentor didn't just console me; he shared his own launch failure story and the specific steps he took to recover. This wasn't just emotionally supportive; it gave me a practical pathway forward that I wouldn't have discovered through coaching alone.

Navigating Unwritten Rules

Every organization has unwritten rules and power dynamics that can make or break a product manager's effectiveness. Mentoring provides a safe space to decode these organizational complexities and develop strategies for navigating them successfully.

When I started mentoring Sharon, a brilliant but frustrated PM, I discovered she was burning political capital by bringing unvetted ideas directly to the CTO. In our mentoring conversations, I shared how I'd learned to build stakeholder support before big meetings and how to read the room during decision-making discussions. These weren't skills she could easily develop through trial and error without significant career damage.

LinkedIn's product org formalizes this aspect of mentoring through their "Organizational Translator" program, where senior leaders help junior PMs understand how decisions really get made, how resources get allocated, and how to build the relationships needed for product success.

Career Development

One of the most valuable aspects of mentoring is providing guidance on long-term career development. As a product leader, you have visibility into career paths and opportunities that your mentees simply can't see from their position. Hopefully, this book will also help with that visibility.

In my mentoring relationships, I make it a point to help PMs identify their unique strengths and how those align with various product career paths. For some, this means advancing toward leadership roles. For others, it might mean specializing in particular domains or product types. The key is helping them make intentional choices rather than simply climbing the next rung on the ladder.

At one company, product leaders take this approach a step further with their "Career Safari" program, where mentors help PMs explore different product roles through temporary assignments and informational interviews before committing to specific career paths.

Creating a Safe Space

Perhaps the most underrated aspect of mentoring is providing a psychologically safe space for product managers to discuss their challenges, doubts, and failures. In an industry that often celebrates success and hides struggle, having permission to be vulnerable is invaluable.

I make it clear to my mentees that our conversations are confidential and judgment-free zones. This allows us to discuss the real issues — like impostor syndrome, conflict with colleagues, or uncertainty about career choices — that they might not feel comfortable raising with their direct managers.

A large e-commerce company's mentorship program institutionalizes this approach with their "Challenge Sessions," where mentees bring their most difficult problems and failures for discussion in a high-trust environment. This normalization of struggle has become a cornerstone of their product culture.

Long-Term Relationships and Commitment

Unlike coaching, which often focuses on specific skills over defined periods, mentoring typically involves longer-term relationships. These extended timelines allow for deeper trust and more comprehensive development.

Some of my most rewarding professional relationships have been with product managers I've mentored over many years. We've navigated multiple role changes, personal challenges, and career decisions together. The longitudinal perspective these relationships provide has been invaluable for both parties.

One popular payment processor takes an interesting approach to this long-term commitment with their "Career Constellation" model, where product managers build a network of mentors who serve different purposes — technical guidance, leadership development, industry perspective — rather than relying on a single mentor for all aspects of growth.

Example: BetterMints

BetterMints, a candy company in Baltimore, transformed their product organization through a thoughtful approach to developing product managers.

Their "Flavor Explorer" program paired junior PMs with senior leaders who helped them map potential career paths based on their unique strengths. When Peter joined as an associate PM, his mentor Shelly didn't just push him toward management — she helped him discover his passion for specialty confections, eventually leading to his specialization in sugar-free innovations.

The company created "Truth Taffy" sessions; monthly gatherings where product managers could discuss their struggles openly. During one session, a senior PM shared how she had completely misread market signals for a major seasonal release. This vulnerability from respected leaders made newer PMs comfortable discussing their own challenges rather than hiding them.

What made BetterMints special was their commitment to long-term development relationships. Unlike other companies with quarterly mentoring rotations, they encouraged connections spanning years. Trisha and her mentor David navigated three role changes together over five years. When Trisha faced a challenging ethical dilemma about sourcing ingredients for her product line, she had the foundation of trust to bring these concerns to David immediately. Their history allowed for nuanced conversations that wouldn't have been possible in a newer relationship.

The result? BetterMints maintained a 92% retention rate for product managers in an industry averaging 70%, while building one of the most respected product organizations in the confectionery sector. And even better: the mentoring program was a template for all other departments across the company.

Balancing Support

Effective mentoring requires a balance of technical skill development and emotional support. Product management is both intellectually and emotionally demanding, and growth requires attention to both dimensions.

A search engine's product mentorship program formalizes this balance through their "Whole PM" approach, ensuring mentoring conversations cover technical skills, emotional intelligence, and overall well-being rather than focusing exclusively on performance metrics.

Requiring Discipline and Accountability

Even the best coaching and mentoring will fall short if they're not paired with appropriate discipline and accountability. This doesn't mean micromanaging or creating punitive environments. Rather, it means establishing clear expectations and helping your team members develop the habits and structures that lead to consistent growth.

When I first started coaching product managers, I would have these wonderful, insightful conversations that led to exciting breakthroughs... that were promptly forgotten amid the daily chaos of product work. I learned that without structured follow-up and accountability mechanisms, even the best development conversations evaporate.

Now I build specific accountability measures into every development relationship. After a coaching session, we document key insights and specific commitments. We schedule follow-up touchpoints to review progress and address obstacles. We create metrics to measure growth in the targeted areas. These structures ensure that development remains a priority rather than an occasional luxury.

Servant Leadership

The foundation of servant leadership is prioritizing your team's needs above your own agenda. This can be particularly challenging for product leaders, who often have strong visions and opinions about product direction.

I've had to learn to distinguish between moments when I should push my own perspective and moments when I should step back to let my team's ideas flourish. Sometimes this means supporting approaches different from what I would have chosen, trusting that the team's ownership and engagement is more valuable than my preferred solution.

At one company, this principle is baked into their planning process. Before product leaders share their own perspectives on roadmap priorities, they first listen to the recommendations from their teams. This sequencing ensures that leaders don't unintentionally override or influence team thinking before it's fully expressed.

Removing Obstacles

One of the most valuable things you can do as a product leader is identify and remove the obstacles preventing your team from performing at their best. These might be organizational barriers, resource constraints, skill gaps, or process inefficiencies.

I spend a significant portion of my one-on-ones asking questions like: "What's getting in your way?" and "Where are you spending time that doesn't create value?" These conversations help me identify the highest-leverage interventions I can make to unblock my team. But don't ask these questions if you're not going to do something to help their situation.

When a PM on my team was struggling with a critical partnership, I recognized that the real obstacle wasn't her skills but her lack of organizational authority. Rather than coaching her on influence tactics (which would have been valuable but insufficient), I used my position to restructure the governance process, giving her a formal seat at the decision-making table. This structural change accomplished more than months of skill development could have.

Amplifying Team Voices

As a product leader, your voice naturally carries weight in meetings and decisions. Servant leadership means using that privilege to amplify your team's voices rather than dominating the conversation.

I've developed several practices to support this principle:

- In meetings, I try to speak last rather than first, ensuring I don't inadvertently set the direction before hearing all perspectives.

- I explicitly credit team members for their ideas, both in their presence and when they're not in the room.

- When presenting to executives, I bring the responsible PMs along and create opportunities for them to showcase their work directly.

Some companies formalize this approach with their "Presenter First" rule; the person who did the work presents it first, with leaders adding context or support only after the team member has had their moment in the spotlight.

Building Consensus Rather Than Commanding

While product leaders sometimes need to make decisive calls, building true consensus creates stronger outcomes and develops your team's decision-making muscles.

I've found that investing time upfront in alignment pays dividends in execution speed and team engagement. This means creating forums where differing perspectives can be openly shared, facilitating structured debates, and helping the team find synthesis among competing viewpoints.

One big online retailer's famous "disagree and commit" principle represents an interesting balance in this area. Teams strive for consensus but acknowledge that complete agreement isn't always possible. When a clear decision is needed, team members commit to supporting the chosen direction even if they initially disagreed.

Empowering Through Delegation and Trust

The ultimate form of development is entrusting your team members with meaningful responsibility and supporting them without taking over. This requires careful matching of challenges to capabilities and setting appropriate guardrails without micromanaging.

When I delegate significant responsibilities — like owning a major product area or leading a cross-functional initiative — I use a tiered approach:

- For newer PMs, I establish more frequent checkpoints and explicitly define the decisions that require consultation.

- For experienced PMs, I focus on clarifying outcomes and principles while giving them significant latitude on approach.

- For my most senior PMs, I primarily serve as a sounding board and barrier-remover, stepping in only when explicitly asked.

A big online music service's product organization has institutionalized this approach with their "Front Stage/Back Stage" model. Leaders stay "back stage" during regular work, providing coaching and support but allowing teams to own their product areas. Leaders only step "front stage" during critical inflection points or when teams explicitly request guidance.

Developing Future Leaders

Perhaps the most important aspect of servant leadership is developing not just strong individual contributors, but future leaders who will carry forward and evolve your leadership approach.

This means deliberately creating leadership opportunities for promising product managers, even before they have formal authority. I regularly identify "leadership moments" where team members can practice skills like facilitating complex discussions, making difficult tradeoffs, or representing the team to senior stakeholders.

For PMs showing leadership potential, I create formal development plans that include shadowing opportunities, incremental leadership responsibilities, and explicit coaching on the mindset shifts required for effective leadership.

Modeling Desired Behaviors

Finally, and perhaps most importantly, servant leadership means modeling the behaviors and mindsets you want to see in your team. Your actions speak far louder than your words or your formal development programs.

If you want thoughtful, user-centered product thinking, you need to visibly practice it yourself. If you want candid communication, you need to demonstrate it in your interactions. If you want appropriate risk-taking, you need to show how you approach and learn from failure.

A streaming video company's product culture exemplifies this principle with their "Context Not Control" philosophy. Leaders focus on creating the context for good decisions through clear strategy, transparent information, and living the values they espouse, rather than controlling specific choices.

Conclusion

While I've discussed coaching, mentoring, and servant leadership as distinct approaches, the most effective product leaders integrate all three in their development work. Different situations and individuals call for different approaches, and sometimes a single interaction might blend elements from each.

The key is developing the self-awareness to recognize which approach is needed in a given moment. Are you helping a PM work through a complex problem they need to solve themselves (coaching)? Are you sharing knowledge from your experience to help them navigate unfamiliar territory (mentoring)? Are you structuring the environment to enable their success (servant leadership)?

I've found it helpful to be explicit with my teams about which hat I'm wearing in different conversations. Starting a discussion with "I'd like to coach you through this decision rather than giving advice" or "I'm going to shift into mentoring mode and share my experience here" helps set appropriate expectations and makes the development process more transparent.

Some of the most successful product organizations have formalized this integrated approach. Intuit, for example, uses their "Design for Delight" coaching program alongside traditional mentoring relationships and servant leadership principles to create a comprehensive development ecosystem.

The Paradox of Product Leadership

The great paradox of product leadership is that your success is no longer measured primarily by what you personally deliver, but by the growth and impact of the people you develop. The truly exceptional product leaders aren't necessarily those with the strongest individual skills or the most impressive shipping record. They're the ones who build organizational capability through others.

This shift can be challenging for many of us who built our careers on personal excellence and direct contribution. However, embracing the disciplines of coaching, mentoring, and servant leadership doesn't diminish your impact; it multiplies it. Every product manager you develop extends your influence far beyond what you could accomplish alone.

I'll leave you with a question that has become my leadership North Star: "Am I building products, or am I building product builders?" The honest answer to this question will tell you whether you're still operating as a super-PM or truly growing into your role as a product leader.

Concepts in Action: ElevationTech

ElevationTech, a boutique software company specializing in computer-generated elevations of building plans, has gained recognition not just for its innovative architectural visualization tools, but for its remarkable approach to developing product talent.

When ElevationTech's founder George Crespin established the company, he brought with him a philosophy shaped by his background in both architecture and software development: "Product managers are like blueprints with different requirements. Some need precise measurements, others benefit from experienced insight, and all require a solid foundation." This perspective led to Elevation-Tech's investment in developing its team with care.

Coaching in Action

Every product manager at ElevationTech participates in the "Blueprint Sessions" program. Unlike traditional performance reviews where leaders simply evaluate work, these bi-weekly meetings focus on expanding thinking through powerful questions.

Technical Director Marcus Lee is known for his coaching approach. In a recent Blueprint Session with Naomi, a product manager struggling with a complex rendering feature, he asked: "What assumptions about architects' workflows are we making that might not align with reality?" After a thoughtful silence, Naomi realized she had been prioritizing speed improvements when architects actually valued accuracy and detail control more highly.

"The insight emerged when I stopped talking," Marcus later explained. "If I had jumped in with my solution, Naomi would have missed discovering a crucial user need herself."

Mentoring Across Experience Levels

ElevationTech's "Construction Knowledge" program pairs product managers with mentors who provide direct guidance based on their experiences. Unlike coaching, these relationships focus on sharing practical insights and navigating both technical complexities and client expectations.

When junior PM Darren joined the visualization team, he was paired with Elena, who had developed similar rendering features years earlier. During their monthly sessions, Elena shared specific strategies that had worked for her, including a story about a failed color palette system that confused architectural clients. This mentoring relationship gave Darren practical wisdom that prevented similar missteps.

The company also maintains "Rendering Journals" where senior developers and product leaders document their major technical decisions, including the context, options considered, and reasons for their final choice. These become invaluable learning tools for developing product managers.

Servant Leadership Philosophy

At ElevationTech, leadership roles come with a clear mandate: "Your foundation supports your team's structure." This principle guides how leaders behave daily.

Director of Product James Whitaker demonstrates this through his "Foundation First" practice. In feature planning meetings, he intentionally speaks last, ensuring team members express their views without being influenced by his position. When a junior PM proposed an unconventional approach to shadow rendering contrary to James's initial thinking, he supported the team's decision rather than overriding it. The feature became one of their most acclaimed by architectural clients.

The company's "Obstacle Removal Tuesdays" institutionalizes the servant leadership concept further. Each Tuesday, leaders clear their calendars to focus exclusively on addressing barriers identified by their teams. After product manager Kai mentioned that coordinating with the engineering team was consuming most of his time, his director created a specialized cross-functional team structure that improved collaboration and reduced meeting overhead by 60%.

Integrated Development in Practice

ElevationTech's most distinctive practice is its "Development Elevations." Just as their software creates different views of a building, product managers work with their leaders to identify which development approach — coaching, mentoring, or servant leadership — provides the best perspective for different growth areas. Visual indicators in their project management system help both parties recognize which "hat" they're wearing in any given conversation.

When Product Lead Sophie needs to develop a team member's client communication skills, she explicitly notes: "I'm going to coach you through this client presentation rather than telling you exactly what to say." This clarity has transformed development conversations throughout the organization.

Measuring Leadership Success

The company's performance review system reflects its development philosophy. Leaders aren't evaluated primarily on the features they ship, but on their teams' growth metrics: How many of their product managers have expanded their technical and business skills? How many have been promoted? How autonomous has their team become?

As CEO George Crespin explains: "At ElevationTech, we're not just building visualization software. We're building the product leaders who will shape the future of architectural rendering. Our greatest competitive advantage isn't our current rendering engine. It's our ability to develop product builders who consistently deliver exceptional work that architects and builders rely on."

Through this integrated approach to development, ElevationTech has created a self-reinforcing culture where coaching, mentoring, and servant leadership combine to nurture the next generation of product leaders in architectural visualization.

Execution Essentials

Coaching Techniques

- Ask powerful open-ended questions that expand thinking rather than narrowing it, such as "What assumptions are you making that might not be true?" or "How would our most demanding customer view this decision?"

- Create psychological safety by sharing your own mistakes first, establishing that it's safe to be imperfect.

- Get comfortable with silence after asking questions; some of the most profound insights come after those seemingly uncomfortable pauses.

- Focus on building problem-solving capabilities rather than solving immediate problems, even when you see the solution clearly.

Effective Mentoring

- Share your own failure stories and recovery steps to provide practical pathways forward for mentees facing similar challenges.

- Help decode organizational complexities and unwritten rules that product managers need to navigate successfully.

- Provide guidance on long-term career development options based on the individual's unique strengths.

- Create a confidential, judgment-free zone where mentees can discuss real issues like impostor syndrome, conflicts, or career uncertainties.

Servant Leadership Principles

- Distinguish between moments to push your own perspective and moments to step back and let your team's ideas flourish.

- Identify and remove obstacles preventing your team from performing at their best, including organizational barriers, resource constraints, or process inefficiencies.

- Amplify team members' voices by speaking last in meetings, explicitly crediting their ideas, and creating opportunities for them to showcase their work directly.

- Build consensus rather than commanding decisions, creating forums where differing perspectives can be openly shared.

- Match delegation challenges to capabilities with appropriate guardrails that evolve as team members grow.

Developing Future Leaders

- Create leadership opportunities for promising product managers even before they have formal authority through "leadership moments."

- Model the behaviors and mindsets you want to see in your team; your actions speak louder than your words or formal development programs.

- Integrate coaching, mentoring, and servant leadership approaches based on each situation's needs.

- Be explicit about which development approach you're using in different conversations to set appropriate expectations.

- Shift your success metrics from personal delivery to the growth and impact of the people you develop.

The Role
Chapter 9

Sustainability &
Ethical Leadership

When I started in product management during the Ice Age, sustainability was that nice-to-have bullet point we'd add to the roadmap when we had extra time (which was never). Ethical considerations? Those were for the legal team to worry about after we'd already built the thing.

Times have changed, thankfully. As product leaders, we now find ourselves at the nexus of some of the most critical challenges facing businesses today. The decisions we make ripple outward, affecting not just quarterly results, but the environment, society, and the long-term viability of our organizations. This chapter isn't about jumping on the corporate social responsibility bandwagon. It's about recognizing that sustainable and ethical product management is simply good business.

Sustainability

Product management has always been about balancing competing priorities. We juggle user needs, business objectives, and technical constraints daily. But a fourth dimension has emerged that demands equal attention: our responsibility to society and the planet. This isn't just idealism talking. Consumers, employees, investors, and regulators increasingly demand that companies consider their broader impact.

Sustainability in product management isn't just about making eco-friendly claims in your marketing materials. It's a comprehensive approach that considers environmental impact throughout a product's existence. Let me break down the key components that have proven essential in my experience.

Lifecycle Analysis

Every product tells a story that extends far beyond what the customer sees. This story begins with raw material extraction and continues through manufacturing, distribution, use, and eventual disposal. Understanding this complete narrative is the foundation of sustainable product management.

I talked extensively with a friend managing a hardware team that was proud of their new smartphone design. On paper, it was 20% more energy-efficient during use than the previous model. Great news, right? But when we conducted a full lifecycle analysis, we discovered that the manufacturing process for the new components generated twice the carbon emissions. The net impact was negative, despite the efficiency gains visible to consumers.

Effective lifecycle analysis requires cross-functional collaboration. Your supply chain, engineering, and operations teams possess crucial insights that marketing and product might miss. One company formed "lifecycle councils" with representatives from each department, meeting quarterly to map impacts and identify improvement opportunities for each product line.

Start by mapping the key stages of your product's life, then quantify the environmental impact at each stage as best you can. Even rough estimates are better than nothing. This will quickly highlight where your biggest opportunities for improvement lie — sometimes in unexpected places.

Circular Design Principles

The traditional product lifecycle is linear: create, use, dispose. Circular design aims to close that loop, keeping resources in use for as long as possible. This approach represents one of the most significant shifts in product thinking in recent decades and one company is excellent at it.

Apple provides an interesting case study here. Their products were once notoriously difficult to repair, with glued components and proprietary screws. But facing consumer pressure and regulatory changes, they've gradually embraced more circular principles. Their latest devices feature more recyclable materials, modular designs for easier repair, and robust trade-in programs. The company now operates one of the world's largest recycling operations for consumer electronics.

Implementing circular design principles starts with asking different questions during product development: How can we design for disassembly? Can components be standardized for easier reuse? Is upgradeability possible without replacing the entire product? These questions often lead to innovations that benefit both customers and the environment.

Stakeholder Alignment

Sustainability initiatives die quickly without proper stakeholder alignment. Your eco-friendly packaging means nothing if the finance department isn't willing to absorb the cost increase, or if manufacturing can't adapt their processes to the new materials.

At a consumer packaged goods company the product team developed a brilliant sustainable packaging solution, only to have it shot down in the final stage because they hadn't involved the operations team early enough. The new packaging would have required significant changes to the production line; changes that weren't budgeted and would have delayed several other products in the pipeline.

Effective alignment requires mapping all stakeholders — internal and external — and understanding their incentives. What matters to your CFO is different from what motivates your customers or concerns regulators. Find the overlap between sustainability goals and each stakeholder's priorities, and frame your initiatives accordingly.

Create an "alignment map" for major sustainability initiatives, which will help navigate potential roadblocks. List each key stakeholder, their primary concerns, what they stand to gain from the initiative, and what they might lose. This exercise often reveals opportunities to adjust your approach for better buy-in while still achieving environmental goals.

Measurable Metrics

"What gets measured gets managed" might be a said too often, but it's very relevant for sustainability initiatives. Not so clear commitments to "being greener" inevitably weaken without specific metrics to track progress and hold teams and companies accountable.

The key is choosing metrics that are both meaningful and practical to measure. Perfect data is hardly available, especially when starting out. Begin with what you can reliably track now, while creating capabilities to measure more sophisticated metrics over time.

Some effective metrics I've seen include: percentage of recyclable materials, carbon footprint per unit, water usage during manufacturing, year-over-year waste reduction, and improvements in energy efficiency. The specific metrics will vary by industry and product type, but should always connect to your broader business objectives.

Supply Chain Transparency

Your product's sustainability profile is only as strong as its weakest link, which is often hidden somewhere in your supply chain. Today's consumer increasingly holds brands accountable not just for their actions, but for the practices of their suppliers and partners.

Patagonia stands out as a leader in supply chain transparency. They've built their brand around it, going beyond industry standards to map their materials back to their sources. Their Footprint Chronicles program publicly documents the journey of their products from raw materials to finished items, highlighting both achievements and areas for improvement.

For most companies, achieving this level of transparency takes time. Start by mapping your tier one suppliers and their environmental practices, then gradually work backward through the chain. Technology is making this process easier, with blockchain and other traceability solutions creating verifiable records of product journeys.

Regulatory Compliance

Environmental regulations are spreading globally, and the pace of change is gaining speed. What's voluntary today may be mandatory tomorrow. Smart product leaders stay ahead of this curve, treating compliance as a minimum baseline rather than a target.

The EU's approach to electronics waste provides a telling example. The original WEEE Directive established basic requirements for electronic waste handling, but subsequent updates have progressively tightened standards. Companies that merely met the initial requirements scrambled to adapt with each revision, while those thinking ahead built more flexible systems that could meet or exceed new and revised regulations.

Maintaining a global regulatory radar is essential for product leaders in multinational companies. I recommend forming a cross-functional team with legal, compliance, and product representation to regularly review emerging regulations and assess their potential impact.

Turn compliance into a competitive advantage by designing flexibility into your products and processes. When California banned certain flame retardants in furniture, one manufacturer had already developed alternatives, giving them a six-month market advantage while competitors reformulated their products.

Green Materials Selection

The materials you choose have cascading effects throughout your product's lifecycle. Each selection represents a trade-off between performance, cost, availability, and environmental impact — a balancing act that defines sustainable product development.

A global carpet manufacturer provides an instructive case study. Under the leadership of sustainability of a pioneer in this space, they transformed their material selection process. Rather than starting with conventional materials and looking for incrementally better alternatives, they established ambitious sustainability criteria and challenged suppliers to meet them. This approach led to breakthrough innovations like carpet tiles made from reclaimed fishing nets.

Creating a materials preference hierarchy can guide your team's decisions without requiring sustainability expertise from every product manager. One company developed a simple system: preferred materials (actively seek opportunities to use), acceptable materials (use with appropriate justification), and materials to avoid (require executive approval for use).

This hierarchy evolved over time as new information became available and alternative materials emerged. The key was establishing a process for regular review and update, ensuring our standards remained both environmentally sound and practically achievable.

Example: PlushLeaf

PlushLeaf, a medium-sized commercial carpet manufacturer, found themselves at a crossroads in 2023. Their stylish office carpets were selling well, but building managers and corporate clients were increasingly demanding transparency about environmental impact. CEO Gregory Best knew they needed to transform their approach.

The company started with supply chain transparency. They mapped their entire supplier network, discovering that 80% of their environmental footprint came from just three materials suppliers in the Midwest. They built a simple dashboard that tracked environmental metrics for each supplier and made it available to clients through a QR code on sample books. Clients could see exactly where their carpet materials came from and their environmental impact. Within six months, sales inquiries increased 30% as architects and designers explored the supply chain data.

For regulatory compliance, PlushLeaf took a proactive approach. Their head of product realized that emerging regulations in Europe would soon affect their US business. Rather than waiting, they formed a small cross-functional team to monitor global regulations and started redesigning products to meet the strictest standards anywhere. When New York passed new rules on VOC emissions in building materials, PlushLeaf was ready while competitors scrambled. Their foresight created a competitive advantage, and major property developers featured them prominently in their green building initiatives.

Their green materials selection process evolved from a frustrating spreadsheet to an innovative supplier partnership program. Instead of simply choosing from available options, they established clear sustainability criteria and challenged suppliers to meet them. One small Georgia fiber mill developed a stain-resistant carpet fiber made from recycled plastic bottles that proved both more sustainable and more durable than their previous material. The "EverCycle" fiber became a cornerstone of their marketing, and carpets using it commanded a 15% price premium that clients willingly paid.

Two years into their transformation, PlushLeaf's sales increased by 23%, and their products earned top sustainability ratings in the commercial building industry. As Gregory put it: "We didn't just become more sustainable — we became a better carpet company."

Energy Efficiency

Energy use represents both an environmental challenge and a direct cost factor for many products. Improvements in this area often create that rare win-win scenario: reducing environmental impact while delivering customer benefits like lower operating costs or longer battery life.

Tesla's approach to energy efficiency demonstrates how this focus can become a core competitive advantage. While other electric vehicle manufacturers were primarily concerned with extending range through bigger batteries, Tesla obsessed over efficiency, from motor design to aerodynamics to regenerative braking. The result was vehicles that traveled farther per kilowatt-hour, reducing both environmental impact and consumer costs.

For digital products, energy efficiency might seem less relevant, but with over 4 billion internet users, the aggregate impact is enormous. Google has led the way here, continuously optimizing their data centers and services to reduce energy consumption. Their efforts have made their data centers 50% more energy-efficient than industry averages.

Customer Education

Even the most sustainable product can have a significant environmental impact if customers don't use, maintain, and eventually dispose of it properly. Effective customer education bridges the gap between your sustainability intentions and actual outcomes.

Seventh Generation, the household product company, excels at customer education. Their packaging and marketing materials clearly explain not just the environmental benefits of their products, but also how proper use amplifies those benefits. For example, their laundry detergent instructions specify cold water washing to reduce energy consumption.

I feel that sustainability information needs to be delivered differently than other product communications. It should be concise, action-oriented, and connected to customer values. A washing machine manufacturer saw much higher adoption of eco-friendly cycles when they relabeled them from technical descriptions ("Eco Mode") to benefit-oriented language ("Energy Saver").

Digital products offer a chance to provide just-in-time education. One streaming service used clever nudges to check if users were still watching after a while without changing channels or programs. If there was no response, it shut down the service, saving on router and server energy.

Innovation Balance

Perhaps the most challenging aspect of sustainable product management is balancing environmental improvements against other product priorities. Push too hard on sustainability at the expense of performance or price, and you risk creating products that customers admire but don't buy.

Unilever's experience with concentrated laundry detergents illustrates this balance. Their first attempts at more sustainable concentrated formulas saved packaging and transportation emissions but required customers to change their dosing habits significantly. Adoption was low despite the environmental benefits. Later versions balanced innovation differently, with clever dosing caps and clearer instructions that maintained user experience while still delivering environmental improvements.

Finding this balance requires a deep understanding of your customers' priorities and willingness to compromise. Not every product can be optimized for sustainability on every dimension simultaneously. Strategic choices about where to focus create more meaningful progress than trying to improve everything at once.

Ethical Leadership in Product

While sustainability focuses primarily on environmental impact, ethical product leadership encompasses a broader set of considerations about how your products affect individuals and society. The two areas overlap significantly but require different approaches.

User-Centered Decision Making

Ethical product management starts with a fundamental commitment: user welfare takes precedence over short-term business gains. This principle sounds obvious until you're staring at quarterly targets with a potential feature that would boost metrics but might not serve users' best interests.

I've witnessed this tension play out repeatedly in the attention economy. One particular social media company faced a decision about a notification feature that significantly increased engagement metrics in testing. However, deeper analysis revealed it was exploiting psychological triggers that users later reported made them feel manipulated. Despite the metrics, they ultimately decided against implementing it.

Facebook's handling of similar decisions provides a cautionary contrast. Their internal research identified potential harms from certain engagement-boosting features, particularly for teenage users. Yet these findings reportedly didn't always lead to changes that might have reduced engagement metrics but improved user welfare.

Building ethical decision-making into your product process requires explicit methods. One approach is a simple checklist of questions for major feature decisions: Does this truly solve a user problem? Are we being transparent about how it works? Could this feature be harmful to vulnerable users? Would we be comfortable explaining our design choices directly to users? How would we feel if the feature affected one of our children?

These questions won't provide perfect answers, but they create valuable pause points for reflection before proceeding with potentially problematic features.

Transparency

Users increasingly expect to understand how products work, how their data is used, and what trade-offs are being made on their behalf. Transparency builds trust and enables informed consent; both essential elements of ethical product management.

A online project management company provides a strong example of transparency in practice. Their product collects workplace activity data that could potentially be misused for surveillance. Instead of downplaying this capability, they're explicit about what data is collected and who can access it. They provide clear controls for users and recommendations to organizations about ethical use of the platform.

Implementing meaningful transparency requires thinking beyond legal requirements like privacy policies that few users read. It means making important information accessible where and when users need it, using plain language and thoughtful design.

Accountability

Accountability means taking ownership of your product's outcomes; both intended and unintended. It requires creating systems that clarify responsibility for decisions and establish consequences for ethical failures.

Google's original approach to AI principles demonstrates how accountability can work in practice. After employee pushback on certain military AI applications, they established clear ethical guidelines for AI development and created a review structure to enforce them. This included appointing executives responsible for implementation and reporting mechanisms for potential violations.

For significant product decisions, record not just what was decided, but why, what alternatives were considered, and who was responsible for the decision. This documentation serves both as an accountability mechanism and as a learning tool when outcomes don't match expectations. Don't go out of your way creating another process. Usually Jira (or whatever software you use to track feature tickets) comments with tags will work just fine. Even if later changed, these comments are contained in the feature's history and for this very reason.

Effective accountability also means acknowledging mistakes quickly and taking corrective action. When one of my teams discovered that a feature was having unintended negative consequences for certain banking customers, we immediately rolled it back out of UAT (user acceptance testing).

Example: TireTracker

TireTracker is a company that created safety software for the trucking industry to monitor the lifecycle of big rig tires. Their system collects tire pressure data, tread depth information, and temperature metrics to help fleet managers prevent blowouts and optimize replacement schedules.

In their early days, TireTracker operated like many tech startups — collect as much data as possible and figure out how to use it later. This approach led to some concerns when fleet managers discovered the system was tracking driver behavior that wasn't directly related to tire safety. Trust began to erode among their trucking company clients.

The turning point came when a new product leader joined the team. She implemented a transparency-first approach that completely transformed how TireTracker handled fleet data. Rather than burying information in lengthy terms of service, they created a simple data dashboard accessible directly in the fleet management portal. This dashboard showed managers exactly what information was being collected, how long it was stored, and who could access it. When driver performance tracking was active, a persistent notification appeared in reports, making it impossible to miss.

TireTracker took accountability seriously by assigning specific team members as "safety stewards" for different features. These stewards were responsible for ensuring their features adhered to the company's ethical guidelines. Their names were attached to feature documentation in Jira so everyone knew who made key decisions. When a feature that shared comparative driver data across different fleets caused unexpected privacy issues for some companies, the responsible team promptly pulled the feature, acknowledged the mistake in a straightforward email to clients, and worked with affected fleets to address concerns.

The team made documentation a priority; recording not just technical specifications but also the reasoning behind decisions and potential safety implications that were discussed. They used Jira to tag important compliance considerations in feature tickets, ensuring this information stayed with the feature throughout development.

These changes had a measurable impact. Client trust metrics improved significantly, and TireTracker saw a 30% decrease in contract cancellations. The team found that being upfront about data practices actually reduced support inquiries and made fleet managers more comfortable sharing information when they understood how it would be used to improve safety.

Inclusive Design

Ethical product leaders recognize that products often fail to serve users equally well across different demographics, abilities, and contexts. Inclusive design aims to address these gaps, creating products that work for the full spectrum of potential users.

Microsoft's evolution in this area offers valuable lessons. Their Inclusive Design toolkit emerged from recognizing that designing for people with permanent disabilities often creates benefits for everyone. For example, features originally designed for users with visual impairments, like high-contrast modes and screen readers, now benefit many users in different contexts.

Implementing inclusive design starts with representation; ensuring your user research includes diverse participants and that your product team reflects diverse perspectives. But it also requires structural changes to your development process, building inclusion checkpoints into your standard workflow.

At a large bank where I worked, we implemented an ADA "inclusion impact assessment" at key stages of product development. This structured evaluation examined how design choices might affect different user groups and identified potential exclusionary elements before they were built.

Privacy by Design

Privacy considerations have shifted from legal compliance concerns to fundamental design parameters. Ethical product leaders build privacy protections into their products from conception, not as afterthoughts.

DuckDuckGo has built its entire business around privacy by design. Their search engine minimizes data collection by default, avoiding the common pattern of collecting everything possible and then managing access. This approach eliminates many potential privacy risks at the source.

Implementing privacy by design requires shifting both mindset and process. Within a bank's resiliency data platform, we replaced the traditional question "What data do we want to collect and store?" with "What is the minimum data needed to deliver this feature and value?" This subtle change led to notably different design decisions and reduced our privacy risk exposure.

Data minimization, purpose limitation, and user control form the foundation of privacy by design. I've found that privacy impact assessments conducted early in the design process highlight risks and opportunities that are much harder to address later in development.

Sustainable Innovation

Ethical product leaders balance rapid innovation with ecosystem health, recognizing that "move fast and break things" can have serious consequences when the "things" are social institutions, mental health, or democratic processes.

Airbnb's evolution shows this balance in action. Their early growth focused on rapid expansion with limited consideration of impacts on housing markets and communities. As problems emerged, they shifted to a more measured approach, working with cities on regulations and implementing features to prevent neighborhood disruption.

Finding this balance means thinking systemically about your product's role in larger contexts. What second-order effects might emerge from your success? How might your product be misused? What dependencies are you creating for users or partners?

One company implements a quarterly "ecosystem health review" alongside their growth metrics. This structured assessment examined how their growth was affecting various stakeholders and helped them identify potential issues before they became crises.

Bias Mitigation

Products inevitably reflect the biases of their creators, their data, and the systems they operate within. Ethical product leaders actively work to identify and address these biases before they cause harm.

A social messaging company's image cropping algorithm provides a cautionary tale. Users discovered that the automated cropping consistently favored white faces over Black faces in preview images. The issue wasn't malicious intent but rather unconscious bias in the algorithm's development and testing that went undetected until public use.

Addressing bias requires both process changes and cultural shifts. Diverse teams are essential but insufficient without structured processes to surface potential biases. An AI company implemented a "bias bounty" program similar to security bounties, rewarding both employees and external researchers for identifying potential biases in their systems.

Regular bias audits should be part of your product development cycle. These reviews examine product decisions, algorithms, and content for potential unfairness across different user groups. The most effective audits involve cross-functional teams and external perspectives to catch blind spots the core team might miss.

Ethical Testing Methods

How we test products can be as ethically significant as what we build. Ethical product leaders develop rigorous methodologies to evaluate potential harms before release, going beyond functionality testing to examine broader impacts.

Netflix's approach to testing recommendation algorithms demonstrates thoughtful ethical testing. Beyond measuring engagement, they assess whether recommendations create filter bubbles or reinforce problematic content patterns, using both quantitative metrics and qualitative evaluation by trained reviewers.

Implementing ethical testing requires expanding both what you test and how you measure success. At a health app company, they developed a method that evaluated features across dimensions including accessibility, potential for misuse, cognitive load, and emotional impact, in addition to standard usability metrics.

The most effective ethical testing happens throughout development, not just before launch. Early-stage concept testing should include ethical dimensions, with increasing rigor as products move toward release. This progressive approach catches potential issues early when they're easier to address.

Purpose-Driven Metrics

The metrics you choose define success for your team, shaping countless decisions and trade-offs. Ethical product leaders move beyond growth and engagement to metrics that reflect meaningful impact and align with organizational values.

Duolingo offers an instructive example. While they track standard engagement metrics, their core success measures focus on learning outcomes; how effectively users are acquiring language skills. This approach aligns their product decisions with their educational mission rather than pure engagement maximization.

Developing purpose-driven metrics starts with clarifying what impact you're truly trying to create. Engagement is rarely an end in itself but rather a means to deliver some more meaningful value. Identifying that core value and measuring it directly creates a more ethical product orientation.

Continuous Ethical Education

The ethical landscape for product development evolves constantly, with new challenges emerging as technology and society change. Ethical product leaders commit to ongoing learning about emerging issues and evolving best practices.

Salesforce's approach to ethical education demonstrates institutional commitment. Their Office of Ethical and Humane Use provides regular training, resources, and consultation to product teams, keeping ethical considerations at the forefront as new technologies emerge.

Building ethical awareness requires both formal education and cultural reinforcement. One particular company established an "ethics champions" network across product teams, with members receiving specialized training and serving as resources for their colleagues. They supplemented this with quarterly ethics workshops addressing emerging topics.

External perspectives are essential for comprehensive ethical education. Engage with experts, critics, and affected communities to understand diverse viewpoints on your product's impact. The most valuable insights often come from those experiencing your product in contexts very different from your development environment.

Conclusion

Sustainable and ethical product management isn't a separate track of work. It's a dimension of all product work. The most effective product leaders integrate these considerations into their standard processes rather than treating them as special cases.

I've found that three key elements enable this integration: leadership commitment, structural support, and cultural reinforcement.

Leadership commitment means demonstrating through decisions and resource allocation that sustainability and ethics are genuine priorities, not just marketing positions.

This includes making difficult trade-offs when necessary and celebrating successes in these areas alongside financial achievements.

Structural support involves building sustainability and ethics checkpoints into your product development process, creating specialist roles where needed, and establishing clear accountability for outcomes. Without these structural elements, good intentions rarely translate to consistent practice.

Cultural reinforcement means nurturing values and behaviors that support sustainable and ethical product development. This includes celebrating team members who raise difficult questions, sharing case studies of both successes and failures, and creating psychological safety for discussing complex trade-offs.

The journey toward more sustainable and ethical product management is neither simple nor short. You'll face complex trade-offs, insufficient information, and competing priorities. But each thoughtful decision moves your products, your organization, and our industry in a better direction. That's leadership worth practicing.

Concepts in Action: InnoEco Solutions

InnoEco Solutions designs and manufactures smart home devices with sustainability and ethics at its core. Founded in 2019 by former environmental engineers, their flagship products include energy-efficient thermostats, water conservation systems, and air quality monitors that help consumers reduce their environmental footprint.

Lifecycle Analysis and Circular Design

"Every product tells a story that goes far beyond what the customer sees," explains Troy Halls, InnoEco's Chief Sustainability Officer. When their product team proposed a new smart thermostat with enhanced AI capabilities, they didn't just focus on its energy-saving potential. Instead, they mapped environmental impacts from raw materials through disposal. This analysis revealed that while the thermostat would save energy during use, the specialized chips would generate substantial emissions during manufacturing.

Rather than accepting this trade-off, Lead Engineer Tom Walker suggested a redesign. "We don't want to just shift environmental impacts from one stage to another," he said during a critical design review. The resulting thermostat features modular construction for easy upgrading and uses recycled plastics from their own take-back program, creating a closed-loop material cycle.

Stakeholder Alignment and Metrics

InnoEco's COO, Jaques Mentel, remembers when sustainability initiatives struggled to gain traction. "Our first water conservation system was technically brilliant but failed in implementation because we hadn't involved operations early enough. Now we create alignment maps for every product." These maps identify all stakeholders from finance to customers, understanding their incentives and finding overlaps between sustainability goals and each stakeholder's priorities.

"What gets measured gets managed," says CFO James Wilson, who helped establish clear metrics for sustainability efforts. Rather than vague commitments to "being greener," InnoEco tracks specific measurements including recycled material percentages, carbon footprint per unit, and customer resource savings.

Supply Chain and Regulatory Compliance

Marketing Director Sara Lopez describes their approach to transparency: "We can't claim to be sustainable if we don't know where our materials come from." Their "Materials Journey" initiative documents the source and impact of major components. When they discovered a supplier was using conflict minerals, Supply Chain Manager Luis Ortiz worked with them to find alternatives. "Cutting them off would have been easier, but collaborative improvement creates more lasting change," he explains.

Beyond compliance, InnoEco's legal team leads a cross-functional regulatory radar group. "Compliance is our floor, not our ceiling," says General Counsel Rachel Kim. This foresight gave them a market advantage when California restricted flame retardants, as they had already developed alternatives months before regulations took effect.

Innovation Balance and Customer Education

"Our first-generation thermostat taught us a hard lesson," admits Product Manager Sue Wilkins. "We prioritized sustainability to such an extent that we created a confusing interface. Customers loved the idea but struggled with the reality." The company's second-generation product maintained strong environmental credentials while significantly improving usability.

UX Designer Emma Johnson works closely with the sustainability team to create intuitive interfaces. "Even the most eco-friendly product fails if people don't use it correctly," she notes. Their packaging and mobile app include clear instructions for optimal use, with intuitive visuals showing how different settings affect resource consumption.

Ethical Product Development

"Technology isn't neutral – it embodies the values of its creators," says Ethics Director Priya Sharma. InnoEco established a rule requiring all features to answer key questions about user problems, transparency, potential harms, and design choices. This rule was tested when data showed behavioral nudges could increase energy savings.

"The marketing team saw a chance to boost our environmental impact metrics," recalls one team member. "But our ethics review identified potential problems for elderly users who might feel uncomfortable with subtle pressures." The final implementation included clear transparency about nudges with simple override options.

CEO Jordan Taylor believes their success stems from integrating sustainability and ethics into everything they do. "We don't make sustainable products," Taylor explains to new employees. "We make excellent products sustainably. That's the difference between greenwashing and genuine leadership." This philosophy has driven both their environmental impact and business success, proving that responsible product management is simply good business.

Execution Essentials

Sustainability Practices

- Conduct comprehensive lifecycle analysis to understand impacts from raw material extraction through disposal, not just during product use.

- Implement circular design principles by creating products that are easier to repair, upgrade, and recycle through modular construction and thoughtful material choices.

- Create alignment maps for major sustainability initiatives to identify each stakeholder's concerns, potential gains, and potential losses.

- Build supply chain transparency beyond tier-one suppliers, gradually mapping your materials back to their source.

Regulatory Approaches

- Form a cross-functional regulatory radar team with legal, compliance, and product representation to regularly review emerging regulations.

- Treat compliance as a minimum baseline rather than a target, staying ahead of regulatory changes.

- Design flexibility into products and processes to adapt quickly to changing regulatory requirements.

- Turn compliance into a competitive advantage by developing alternatives before regulations take effect.

Material Selection

- Create a materials preference hierarchy to guide decisions: preferred materials, acceptable materials, and materials to avoid.

- Establish a process for regular review and update of material standards as new information and alternatives emerge.

- Challenge suppliers to meet ambitious sustainability criteria rather than only seeking incrementally better alternatives.

- Consider impacts across the entire lifecycle when selecting materials, not just during manufacturing or use.

Energy Efficiency

- Focus on energy efficiency as a core competitive advantage, not just an environmental benefit.

- Identify efficiency improvements that create win-win scenarios for both environment and customers.

- Consider the aggregate impact of digital products used at scale, not just individual device efficiency.

- Design products to help consumers reduce their energy consumption through smart features and settings.

- Optimize for efficiency across all components, not just the most visible ones.

Ethical Leadership

- Make user welfare a priority over short-term business gains when making product decisions.

- Ensure user research includes diverse participants and that your product team reflects diverse perspectives.

- Recognize that designing for people with permanent disabilities often creates benefits for everyone.

- Test products with users from different demographics, abilities, and contexts.

- Replace "What data do we want to collect?" with "What is the minimum data needed to deliver value?"

- Build privacy protections into products from conception, not as afterthoughts.

- Make important privacy information accessible where and when users need it, using plain language.

Leadership Integration

- Demonstrate leadership commitment through decisions and resource allocation that prioritize sustainability and ethics.

- Build checkpoints for sustainability and ethics into your standard product development process.

- Create specialist roles where needed and establish clear accountability for outcomes.

- Nurture values and behaviors that support sustainable and ethical product development.

- Celebrate team members who raise difficult questions and create psychological safety for discussing complex trade-offs.

Part 2

The Product

The Product

Chapter 10

Product Alignment & Business Strategy

I've learned one plain truth. No product (or organization) exists in a vacuum. The best product in the world fails when it doesn't align with business objectives or lacks ethical sustainability. I've witnessed brilliant innovations collapse when they chased market fit without considering long-term viability.

Product management isn't just about shipping features or hitting sprint goals. It's about creating sustainable value; sustainable for your business economics, sustainable for your organization's resources, and sustainable for our shared world. This chapter explores how product leaders align product strategy with broader business goals. Let's get started.

Strategic Vision

Alignment begins with clarity. Without a crystal-clear vision of where you're headed, even the most talented teams will build impressive solutions for the wrong problems. Think of strategic vision alignment as the foundation upon which everything else is built — get this wrong, and no amount of execution excellence can save you.

I've seen billion-dollar companies toppled by startups not because they executed poorly, but because their strategic vision lost relevance while they were busy optimizing yesterday's metrics. The following methods will help you establish that crucial foundation of alignment between your product direction and your company's ultimate destination.

Finding the North Star

In my early days, I fell into the trap many do; measuring everything because we could. We tracked 47 different metrics for a web application that, in reality, only needed a few. We drowned in data while starving for insight. The result? A scattered roadmap that satisfied no stakeholders and confused our dev team.

North Star metrics aren't just convenient measurement tools; they're decision-making methods that align your entire organization. When Netflix shifted from a DVD rental service to a streaming platform, they could have chosen dozens of metrics to track: catalog size, device coverage, or technical performance. Instead, they famously rallied around one primary metric: viewing hours. This single focus guided thousands of micro-decisions across their organization. Product features, content acquisition, UI design; every decision could be evaluated against "will this increase total viewing hours?"

Your North Star metric must tie directly to your business value creation model. For subscription services, it might be retention rate; for marketplaces, transaction frequency; for content platforms, engagement depth. The power comes not from the metric itself but from the clarity it provides. When my teams know our North Star metric, they make better autonomous decisions. They can evaluate their own ideas against this method without waiting for my approval on every minor feature tweak.

Aligning to Company Mission

"We want to organize the world's information and make it universally accessible and useful." Do you recognize Google's mission statement? Their product decisions filter through this lens; from search algorithms to knowledge panels to archiving books. When Google ventured into social networking with Google+, the project ultimately failed. In my opinion, it was partly because it struggled to connect back to this core mission in a meaningful way.

I've found that teams working on products with clear mission alignment have much higher retention rates than those working on products with strategic misalignment. The reason isn't mysterious. People want to work on something they believe matters. When the product they're building clearly serves a mission they believe in, motivation is easily measured.

Fitting to a company's mission requires constant vigilance, especially during growth phases. A healthtech company faced a lucrative opportunity to expand into fitness tracking; seemingly adjacent to their health monitoring roots. However, after mapping the new features against their mission of "improving clinical outcomes through continuous monitoring," they realized they would be diluting their focus. They passed on the opportunity, which initially disappointed investors but ultimately strengthened their position in their core market.

Setting your company's mission early, clearly, and as a guiding principle for your team is crucial. It's not set in stone and can evolve, but it should be a stable foundation. If it does change, it should be done sparingly and ensure that all your current products align perfectly with the new vision.

Clarifying Value Proposition

I once inherited a product with fifteen different value propositions, depending on which team member you asked. The sales team focused on cost savings. Marketing emphasized innovation. The founder talked about disruption. Our users, meanwhile, primarily valued the time savings. No wonder our conversion rates stagnated! We were solving a different problem in every communication.

Your value proposition isn't just marketing fluff. It's the translation of your product's features into customer benefits that acknowledges the specific pain you're removing. Strong value propositions speak to both rational and emotional needs. They explain why your solution is meaningfully different from alternatives, including the alternative of doing nothing at all.

Slack didn't just build a chat tool; they promised a substantial decrease in internal email. They didn't just offer team communication; they offered fewer meetings and more productive teams. These specific value propositions guided their product development, growth tactics, and retention strategies. When you align your product decisions with a clear value proposition, feature prioritization becomes more straightforward. Every potential enhancement can be evaluated against "does this strengthen our core value proposition?"

Competitive Positioning

The graveyard of failed products is filled with "better" solutions that failed to carve out distinctive market positions. One company learned this lesson painfully with a project management tool that was technically superior to market leaders but struggled to find adoption because it was positioned as "Jira, but better." The developer had built a faster horse when the market was ready for automobiles.

Competitive positioning isn't about being better; it's about being different in ways that matter to a specific customer segment. Southwest Airlines didn't compete with United on routes or service quality; they competed on simplicity and price. They didn't try to be a better version of existing airlines; they created a different category.

Your product positioning should focus on the dimensions where you can win. After studying the project management market more carefully, the company above repositioned their tool specifically for creative agencies with complex client approval workflows — an underserved niche where their unique features addressed specific pain points. This narrow but deep positioning allowed them to capture a massive share of this segment within two years, compared to their previous low penetration in the general market.

Long-term Roadmap Planning

The tension between short-term wins and long-term vision creates more product strategy casualties than perhaps any other factor. I've watched quarter-focused thinking transform promising products into POS (piece of stink) features that please no one. The most successful product leaders I've worked with maintain a rolling three-horizon roadmap that connects immediate deliverables to medium-term opportunities and long-term vision.

Long-term roadmap planning isn't about predicting the future; it's about creating enough strategic clarity that tactical decisions become easier. When Amazon invested in AWS, they weren't responding to quarterly pressures. They were building infrastructure aligned with Bezos' vision of Amazon as a technology company, not just a retailer. This long-term thinking allowed them to create an entirely new business line that eventually generated more operating income than their retail operations.

In my experience, effective long-term roadmaps balance specificity and flexibility. At one company, we maintained an 18-month roadmap with decreasing specificity over time. The first six months contained specific features with delivery dates. The next six months outlined capability areas we intended to develop. The final six months focused on customer problems we planned to solve, without committing to specific solutions. This approach gave our development teams tactical clarity while preserving strategic flexibility as market conditions evolved.

Example: Peak Health

Peak Health started as a small health app trying to do everything at once. Their dashboard had 30+ metrics and user growth stalled at 20,000 monthly users.

During a leadership retreat, CEO Sarah Bauer recognized their lack of strategic clarity. They established "meaningful health improvements tracked by users" as their North Star metric, which helped everyone evaluate their work against a common standard.

They refined their mission to focus on personalized health insights, clarified their value proposition as "Personal health insights that doctors actually use," and positioned themselves as a specialized platform for people who needed to share reliable data with healthcare providers.

Their three-horizon roadmap aligned immediate EHR integration features, mid-term health pattern recognition, and long-term AI coaching with their mission.

Within 18 months, Peak Health grew to 350,000 monthly users and secured partnerships with three major hospital networks, transforming from a struggling app into a focused platform creating real value for their specific user segment.

Market & Customer Alignment

Even the most brilliant strategy fails when it doesn't connect with market realities and genuine customer needs. There have ben countless products built on clever strategies that flopped spectacularly when they met actual customers.

The market doesn't care about your elegant positioning or your executive's pet features. It rewards solutions that solve real problems better than alternatives.

Solutions for Real Customer Problems

The most elegant solution to the wrong problem still fails. Early in my product management career, I invested six months building an advanced analytics dashboard that offered twenty different visualization options. The engineering was impressive. The UX was intuitive. Yet adoption remained abysmal. When we finally conducted proper customer interviews, we discovered our users didn't want more visualization options. They wanted pre-configured reports that answered specific business questions.

Problem-solution fit begins with problem verification. I've adopted a practice called "problem before prototype," where product teams must document evidence of the problem's existence, scope, and impact before any solution discussions begin. This evidence might come from support tickets, customer interviews, usage data, or market research, but it must extend beyond anecdotes or assumptions. Typically all of this good data is attached as artifacts to a Jira epic or story.

The payments platform Stripe grew rapidly not because they offered a fundamentally different service than competitors, but because they recognized and solved the actual developer problem: implementation complexity. By focusing on the real friction point (integration time rather than transaction fees), they aligned their solution with the genuine market need. Their seven lines of code integration wasn't just a technical achievement; it was evidence of deep problem understanding.

Market Timing Analysis

Market timing can be a wrecking ball if you're not careful. Teams have built technically excellent products that failed because they were too early (a mobile payment solution in 2008) and others that failed because they were too late (a project management tool in 2019). The right solution at the wrong time is still the wrong solution.

Timing analysis requires both market readiness assessment and competitive landscape evaluation. Market readiness encompasses technological enablers (are supporting technologies mature enough?), regulatory environment (do legal rules support or hinder adoption?), and customer mindset (are customers mentally ready for this change?). The competitive landscape timing considers whitespace identification (are there unaddressed needs?) and differentiation opportunities (can we meaningfully distinguish ourselves?).

Electric vehicle maker Tesla provides an instructive example of timing analysis. While electric vehicles had been attempted for decades, Tesla launched when lithium-ion battery density reached sufficient levels, environmental concerns were heightening, luxury car buyers were seeking differentiation, and regulatory incentives created favorable economics. Their timing wasn't accidental; it was a calculated assessment of multiple converging factors.

Customer Journey Mapping

Products don't exist in isolation; they exist within customer workflows and life contexts. An educational product failure involved a sophisticated task management system that solved all the problems users told about, yet saw minimal adoption. Through contextual inquiry, it was discovered that while the solution addressed users' task management problems in theory, it failed to integrate into their actual daily workflow that involved constant context switching between email, meetings, and documents.

Effective journey mapping goes beyond simplistic funnel analysis to understand the emotional, situational, and environmental contexts of product usage. It answers questions like: What was the user doing immediately before using our product? What other tools or processes are simultaneously in use? What emotional state are they in when engaging with our solution?

Intuit's TurboTax succeeds not because it performs calculations competitors can't match, but because it deeply understands the tax preparation journey. They recognize that users approach tax preparation with anxiety, limited financial knowledge, and fear of mistakes. Their entire experience design — from reassuring language to knowledge validation — addresses these contextual factors rather than merely focusing on the functional task of calculating taxes.

Journey maps are a fantastic way to uncover hidden pain points. Start by creating a step-by-step diagram (with screenshots or images) on a timeline and of each action a user takes when performing a specific task. Tag each step with a green, yellow, or red star. Then, use a spreadsheet filter and sort the steps. Greens are great, yellows might need some improvement, and reds definitely need a better solution. In short time, you might drastically improve your product's usability. Here's some tips:

- Start with clear objectives — define what you want to learn before mapping
- Use real customer data rather than assumptions whenever possible
- Include emotional touchpoints, not just functional interactions
- Map the "as-is" journey before designing the "to-be" journey
- Involve cross-functional teams to gain diverse perspectives
- Look for pain points and moments of delight equally
- Consider both digital and physical touchpoints
- Create personas based on behavioral patterns, not demographics alone
- Validate your maps with actual customers
- Keep maps living documents that evolve with new insights

This book is not meant to be illustrative and creating journey maps could be a book in itself (lightbulb moment?). There is plenty of software and instructional videos at your fingertips, specifically for customer journey mapping:

- Opt for a collaborative canvas, such as Figma, Canva, or Miro. Each of these are online, team-oriented, and very inexpensive.
- Using YouTube, search for "Customer Journey Maps" to find a load of great examples; even ones using the sites above.

- Avoid complicated grids of text and opt for journeys which go from left to right on a timeline – showing illustrations, describing user emotions, and flagging for improvements.

Segmenting Target Users

The myth of the universal user has led countless products astray. One startup proudly declared their product was "for everyone." This lack of focus led to a scattered feature set that served no one particularly well. Their growth stalled until they narrowed their focus to creative hair professionals in independently owned salons, which allowed them to tailor their solution to their specific needs.

Effective segmentation goes beyond basic demographics to understand behavioral patterns, contextual needs, and psychographic characteristics. The goal isn't to exclude potential customers but to focus limited resources on the segments where you can create disproportionate value and establish early success.

Airbnb's early growth strategy exemplifies the power of precise segmentation. Rather than targeting all travelers, they initially focused on attendees of design conferences and events where hotel capacity was constrained. This narrow focus allowed them to tailor their early product and marketing to serve a specific use case exceptionally well before gradually expanding to adjacent segments.

Voice of Customer Feedback

Customer feedback integration presents a classic product management paradox: customers rarely know what solutions they need, but they always know what problems they have. I've seen product teams fail on both extremes — some building exactly what customers requested without interpretation (resulting in feature bloat), others ignoring customer input entirely in favor of their "vision" (resulting in market disconnection).

The most common mistake I encounter is treating customer feedback as a democracy where the most requested feature automatically wins. Early in my career, I fell into this trap with a project management tool. Our feedback portal showed "Gantt chart support" as the most upvoted feature by a significant margin. We spent three months building a sophisticated Gantt implementation only to discover that while many users thought they wanted this feature, fewer than 8% actually used it after release. What they actually needed was better visibility into project dependencies, which a Gantt chart addresses in only one specific way.

Successful voice of customer integration requires triangulation between explicit requests ("I want feature X"), implicit needs revealed through behavior analysis ("users are creating workarounds for problem Y"), and strategic interpretation ("what they're really asking for is outcome Z"). This triangulation transforms raw feedback into actionable insight.

Try to employ a five-step process for meaningful customer feedback integration:

1. Segment feedback by user type and behavior patterns. Feedback from power users who spend four hours daily in your product carries different weight than casual users who engage monthly.

2. Look for underlying jobs-to-be-done rather than focusing on feature requests. When a customer says "I need to export my data to Excel," they're often really saying "I need to analyze this data in ways your interface doesn't support."

3. Validate patterns through quantitative usage data. If customers ask for feature X but behavioral data shows they rarely use similar capabilities, deeper investigation is warranted.

4. Prototype before building. When we received consistent requests for a "dashboard customization" feature, rather than immediately committing development resources, we created clickable mockups of three different approaches and tested them with requesting customers. Their interaction with the prototypes revealed needs that none of our initial designs fully addressed.

5. Close the feedback loop by communicating with customers about how their input influenced product decisions, even when the outcome wasn't exactly what they requested.

At Salesforce, product teams categorize customer feedback into "problem reports" rather than "feature requests," deliberately reframing suggestions to focus on the underlying need rather than the proposed solution. This approach allows them to address the core issue while potentially finding more elegant solutions than what was specifically requested. Their success demonstrates that respecting customer voice doesn't mean implementing every suggestion, but rather deeply understanding the problems behind those suggestions.

The most sophisticated voice of customer integration I've witnessed came from Spotify's product team. They developed a dedicated "insights pipeline" that combined explicit feedback (surveys, support tickets), implicit signals (behavioral analytics, search patterns), and contextual research (in-home observation studies). This multi-channel approach helped them identify that while users explicitly requested the ability to "see what friends are listening to" (a social feature), their underlying need was actually for personalized discovery of new music. This insight led to their enormously successful Discover Weekly feature, which addressed the core need for discovery without the privacy concerns and implementation challenges of a purely social approach.

Business Performance Metrics

The hard truth I've learned in product leadership: customer love doesn't pay the bills. I've built products with passionate user bases that still failed because they couldn't generate sustainable economics. Every product decision has financial implications, whether immediate or long-term.

The most successful product leaders I know have developed a financial fluency that lets them translate user metrics into business outcomes. They don't leave financial analysis to the finance department; they build it into their product thinking.

Understanding the business performance metrics in this section isn't about becoming an accountant. It's about gaining the credibility and tools to advocate effectively for your product in an organization where resources are always constrained and ROI (Return On Investment) always matters.

Forecasting Revenue

I've pitched many product proposals to executives. The ones that received approval weren't necessarily the most innovative or elegant. They were the ones with the most credible connection to revenue growth. While product managers often focus on user metrics, executives ultimately evaluate investments based on financial impact.

Revenue impact projections require translating product outcomes into business outcomes. This translation isn't about manufacturing unrealistic financial forecasts; it's about creating logical connections between product metrics and business metrics. If your product will improve conversion rates by 0.5%, what does that mean for quarterly revenue? If your feature reduces churn by 3%, how does that affect customer lifetime value?

Adobe's transformation from perpetual licensing to subscription services represents a masterclass in revenue impact analysis. They didn't just change their business model; they methodically projected how subscription pricing would affect near-term revenue decline, long-term revenue stability, customer acquisition costs, and lifetime value (LTV). This comprehensive financial modeling gave them confidence to weather the initial revenue dip for long-term financial health.

Cost-Benefit Analysis

Not all features are created equal; either in value delivered or resources required. One company championed a complex accounting software integration feature that customers frequently requested. After three months of development and considerable technical debt, they discovered that only 2% of users activated the feature, and most abandoned it after one use. They had failed to conduct a rigorous cost-benefit analysis before committing resources.

Effective cost-benefit analysis extends beyond development hours to consider ongoing maintenance costs, technical debt implications, opportunity costs, and support requirements. It weighs these costs against projected benefits including revenue impact, competitive differentiation, and strategic positioning value.

The streaming service Netflix exemplifies disciplined cost-benefit analysis in their feature development. Despite user requests for offline viewing capabilities for years, they resisted implementation until both the cost side (technical challenges, licensing complications) and benefit side (competitive pressure, user retention impact) of the equation made sense. This calculated patience allowed them to introduce the feature at the optimal time with proper execution rather than rushing a suboptimal implementation.

Resource Allocation Finances

Resource allocation may be the most political aspect of product management. I've witnessed fierce battles over engineering headcount between product lines, with decisions often made based on executive relationship strength rather than business impact potential. This approach inevitably leads to underinvestment in high-potential areas and over-investment in legacy products.

Optimized resource allocation requires portfolio thinking; viewing your products and features as investment opportunities with different risk-return profiles. Just as financial advisors recommend diverse investment portfolios, product organizations should balance resources between sure bets (incremental improvements to proven products), calculated risks (new features with strong evidence of demand), and strategic options (exploratory initiatives that could unlock new growth vectors).

Microsoft's revival under Satya Nadella demonstrates strategic resource reallocation at scale. By shifting significant resources from Windows to cloud services like Azure, they transformed their growth trajectory and market position. This wasn't just about following market trends; it was a calculated reallocation based on projected return on investment across their product portfolio.

Determining Return on Investment (ROI)

After leading dozens of product development initiatives, I've observed that development efficiency varies by at least an order of magnitude between teams — even those with similar talent levels. Some teams create massive customer value with modest resources, while others consume substantial resources with minimal market impact. The difference often comes down to how rigorously we measure and optimize return on product investment.

ROPI (Return on Product Investment) analysis examines the value created relative to resources invested. This requires defining clear success metrics before development begins, tracking both resource consumption and value creation throughout the development lifecycle, and conducting honest retrospectives about efficiency.

Facebook's "move fast and break things" era provides an instructive counterpoint on development efficiency. While their velocity was impressive, they eventually discovered that technical debt and quality issues were reducing their overall ROPI. Their subsequent shift to "move fast with stable infrastructure" reflected a more sophisticated understanding that pure development speed doesn't necessarily maximize return on investment when considering the full product lifecycle.

Pricing Strategy

Pricing is product management's most powerful lever, yet many product managers abdicate pricing decisions to marketing or sales functions. This disconnection leads to pricing structures that fail to align with how value is actually delivered and perceived. I've seen extraordinary products struggle financially due to pricing models that didn't match their value creation patterns.

Effective pricing strategy goes beyond competitive benchmarking to deeply understand value attribution (which features or outcomes do customers value most?), usage patterns (how does consumption behavior vary across segments?), and psychological factors (how do customers perceive price relative to alternatives, including doing nothing).

Slack's pricing evolution demonstrates sophisticated alignment between product experience and pricing model. Their per-active-user approach precisely matched their value delivery pattern, where benefits increased with broader adoption within an organization. This alignment created natural expansion revenue as product usage grew, without friction-generating renegotiations. Their free tier design strategically limited searchable message history; preserving the core experience while creating a natural upgrade trigger as usage deepened.

An entire chapter in this book is dedicated to monetization and pricing strategy.

Example: InvoCat

InvoCat started as a plucky software tool for small businesses to manage their inventory. The founding team — Amy, Miguel, and Jordan — had built a solid product, but their business metrics were all over the place. They were charging a flat monthly fee, had no idea if they were making money, and couldn't figure out which features were worth investing in.

Then came their turning point. After a particularly painful quarter where they nearly ran out of cash, the team decided to get serious about their business performance metrics.

First, they tackled revenue forecasting. They moved from their one-size-fits-all pricing to a tiered model based on inventory volume. The team used basic cohort analysis; grouping different types of small businesses inventory keepers into separate metrics. They were then able to project how different pricing scenarios would affect their cash flow over the next 18 months — showing them they could double revenue with some customer groups within a year if they executed properly.

The cost-benefit analysis came next. Instead of building every feature customers requested, they started analyzing development costs against projected usage and revenue impact. When enterprise customers asked for advanced reporting tools, InvoCat did the math: 80 developer days at an internal cost of $60,000, but with potential to unlock $400,000 in annual revenue from high-value accounts — now you're speakin' my language!

For resource allocation, they instituted a quarterly portfolio review. Features and products were categorized as "cash cows" (core inventory tools), "rising stars" (their new supplier management module), and "exploratory bets" (blockchain verification they were testing). This helped them distribute developer time proportionally instead of chasing every shiny penny.

Their ROI calculations became increasingly sophisticated. They tracked how many developer hours went into each feature and measured the resulting impact on user acquisition, retention, and expansion revenue. This revealed their notification system had generated a 12x return on investment, while their analytics dashboard barely broke even.

Finally, they overhauled their pricing strategy based on actual usage data. They discovered customers valued certain features far more than expected; particularly batch processing and supply chain integration. This led them to unbundle these high-value components as premium add-ons, increasing average revenue per user by 37% while actually reducing churn.

Three years after implementing these business performance metrics, InvoCat was acquired for $28 million — a far cry from the struggling startup that was nearly thrown out onto the streets.

Organizational Synergy

No product is an island, and no product leader succeeds alone. After years of painful lessons, I've come to realize that organizational alignment is often the hidden multiplier that separates wildly successful products from mediocre ones.

I've built technically superior products that failed because marketing couldn't message them effectively, sales couldn't sell them confidently, and operations couldn't support them adequately. The harsh reality of product leadership is that great product decisions are necessary but insufficient. You need the entire organizational machine working in harmony to deliver exceptional outcomes.

Cross-Functional Collaboration

The greatest product strategies fail when executed in organizational silos. Early in my career, I proudly delivered a product on schedule and with all specified features, only to discover that sales wasn't prepared to sell it, marketing had no campaign ready, and support hadn't been trained. The product technically launched, but with minimal market impact.

Effective cross-functional collaboration requires structural enablement (the right forums and processes), cultural reinforcement (rewards for collaborative behavior), and leadership modeling (executives demonstrating collaborative decision-making). Product managers serve as the connective tissue in this ecosystem, translating between different functional languages and priorities.

Apple's product development process remains the gold standard for cross-functional collaboration. Their "directly responsible individual" (DRI) approach ensures clear ownership while their design reviews bring together perspectives from hardware, software, marketing, and manufacturing early in the process. This integrated approach is why their products feel cohesive across all dimensions: the physical device, operating system, marketing message, and retail experience all tell a consistent story.

I have written an entire chapter about cross-functional team management earlier in this book.

Resource Capacity Planning

Ambitious roadmaps colliding with limited resources create predictable disappointment. In one organization, the team consistently planned for 30-40% more work than their teams could realistically deliver. This created a culture of perpetual shortfall, where even extraordinary performance felt insufficient. Team morale suffered, and stakeholder trust eroded as commitments consistently went unmet.

Effective capacity planning begins with honest assessment of true capacity, accounting for maintenance work, technical debt, employee development time, and inevitable unexpected issues. Against this realistic capacity baseline, prioritization becomes a transparent exercise in trade-offs rather than a futile attempt to accommodate every request.

Toyota's production system offers valuable lessons for product development capacity planning. Their focus on eliminating "muda" (waste) and maintaining "heijunka" (workload leveling) recognizes that overloading systems reduces overall throughput and quality.

When one firm applied similar principles to our product development process — limiting work-in-progress and establishing consistent delivery cadences — they increased their output significantly while improving quality metrics. That also allowed them to better plan future products and features.

Objectives & Key Results (OKRs)

Product initiatives disconnected from company objectives become vulnerable during resource allocation discussions and organizational priority shifts. I've seen promising product lines abruptly defunded when executives couldn't clearly connect their outcomes to broader company priorities.

OKR (Objectives and Key Results) alignment creates a clear traceability from company-level goals through department objectives to product initiatives. This alignment isn't a one-time mapping exercise; it requires ongoing recalibration as both company objectives and market conditions evolve.

Intel's famous adoption of OKRs under Andy Grove demonstrates how this method creates organizational coherence around product decisions. Their approach cascaded corporate objectives into team-level key results, creating line-of-sight from individual product decisions to company-level outcomes. This alignment allowed them to make dramatic shifts in product strategy, including the famous "right-hand turn" away from memory chips toward microprocessors, with organizational cohesion rather than fragmentation.

Portfolio Balancing

As organizations grow, product portfolio management becomes increasingly complex. In one technology company, they expanded from two core products to fifteen in just three years. Without intentional portfolio management, they found themselves with overlapping capabilities, inconsistent customer experiences, and internal competition for resources. Revenue grew, but profitability suffered.

Effective portfolio management requires clear product categorization (core products, growth bets, maintenance mode offerings), deliberate cannibalization strategies (how new offerings should replace older ones), and portfolio-wide investment allocation methods (how to distribute resources across the portfolio).

Procter & Gamble's approach to product portfolio management offers valuable lessons. Their clear distinction between managing established brands and incubating new offerings prevented their innovation initiatives from being suffocated by short-term performance pressures on existing products. This separation allowed them to simultaneously optimize mature product lines while cultivating new growth opportunities.

Stakeholder Expectation Management

Stakeholder expectations may be the most underestimated factor in product success. I've delivered products that met all specified requirements yet were considered failures because stakeholder expectations had evolved without corresponding adjustments to the product plan. Conversely, I've shipped products with significant compromises that were celebrated as successes because expectations had been carefully managed throughout development.

Effective expectation management involves stakeholder mapping (identifying who will judge success), regular calibration (confirming that expectations remain aligned with reality), and proactive adjustment (resetting expectations when conditions change). This isn't about lowering the bar; it's about ensuring the bar is consistently visible and achievable.

Amazon's approach to shareholder expectation management offers a valuable model for product leaders. Jeff Bezos famously told shareholders that Amazon would prioritize long-term value creation over short-term profits, establishing expectations that allowed them to make substantial investments in AWS, Prime, and logistics infrastructure that initially depressed earnings but created enormous long-term value. This stakeholder alignment gave them the runway to build market-defining products that required multi-year investment horizons.

In an earlier chapter, I write fully about stakeholder management.

Example: StreamIndie

StreamIndie, a video streaming platform for independent filmmakers, was struggling to gain traction despite having solid technology. Their new product leader, Hannah Hawke, realized their issues stemmed from organizational misalignment rather than product flaws.

When Hannah joined, the engineering team was building features in isolation, marketing was crafting messages that didn't match the actual product capabilities, and sales reps were making promises the platform couldn't fulfill. Customer support was often blindsided by new releases, leaving them unprepared to help frustrated users.

Hannah's first move was establishing weekly cross-functional standups — bringing together product, engineering, marketing, sales, and support teams to share updates and align priorities. These meetings broke down communication barriers and created a shared understanding of what they were building and why.

For resource capacity planning, Hannah implemented a more realistic approach. Instead of cramming every feature request into each sprint, they limited work-in-progress (each developer could have 2 WIP items in progress) and built in time for maintenance and technical debt reduction. This created predictable delivery cycles and improved team morale as they began consistently meeting commitments.

StreamIndie adopted OKRs (Objectives and Key Results) to connect individual team efforts to company goals. Each quarter, they set clear objectives like "Increase filmmaker retention by improving upload experience" with measurable results such as "Reduce upload failures by 30%." This framework gave everyone clarity on how their work contributed to company success.

As StreamIndie expanded their offerings to include both streaming and promotional tools, they needed thoughtful portfolio management. Hannah created clear categorization: their streaming platform as the core product, analytics as a growth bet, and their older desktop uploader in maintenance mode. This helped them intelligently allocate resources rather than spreading themselves too thin.

The final piece was stakeholder expectation management. Hannah established regular executive reviews where actual progress was transparently shared — both wins and challenges. When they needed to delay a major feature, Hannah proactively reset expectations with key stakeholders, explaining the trade-offs and long-term benefits of the decision.

Within a year, StreamIndie transformed from a technically impressive but under-performing platform to a cohesive product ecosystem with strong market growth. Their success wasn't from radical technology changes but from creating organizational synergy that aligned everyone around delivering genuine customer value.

Execution & Operation

Strategy without execution is hallucination. I've sat through countless product strategy sessions where brilliant visions were articulated, only to watch them collapse when they collided with implementation realities.

The truth that many product leaders don't want to hear: execution quality often matters more than strategy sophistication. The best product strategy poorly executed fails every time, while even a middling strategy executed with excellence can succeed. Too many product leaders focus on the "what" and "why" while neglecting the crucial "how."

Measuring Development Velocity

Development speed creates strategic optionality. Teams that reliably deliver working software every two weeks have many opportunities during the year to learn from market feedback. Teams on quarterly release cycles have just four learning opportunities per year. This difference fundamentally changes how quickly products can evolve toward product-market fit.

Effective velocity metrics focus on outcomes rather than activities. Instead of tracking story points completed (an activity metric), measure working features delivered (an outcome metric). Instead of tracking hours worked, measure cycle time from concept to customer availability. These outcome-oriented metrics prevent the illusion of productivity without actual progress.

Etsy's engineering organization demonstrates sophisticated velocity measurement. Rather than simply tracking deploys per day (a vanity metric given their continuous deployment model), they measure time from commit to production and learning cycle time (how quickly customer feedback influences product changes). This focus on learning velocity rather than just shipping velocity has enabled them to rapidly evolve their marketplace features based on actual user behavior.

Quality Assurance Standards

Quality isn't just about bugs; it's about the gap between customer expectations and experience. I've seen technically flawless products fail because they didn't meet unstated customer expectations about performance, usability, or integration capabilities. Conversely, I've seen products with known technical issues succeed because they delivered extraordinary value in dimensions customers cared about most.

Comprehensive quality assurance encompasses functional correctness (does it work as specified?), performance (does it work efficiently?), usability (is it intuitive?), reliability (does it work consistently?), and security (is it protected against threats?). The relative importance of these dimensions varies by product context. Medical devices prioritize reliability and security, while consumer apps may emphasize usability and performance.

Toyota's "quality first" approach to manufacturing offers valuable lessons for product development. Their Andon cord system, which allows any worker to pull a cord (most likely push a button) and stop production when quality issues are detected, inspired modern DevOps practices like automated testing and continuous integration. Both approaches recognize that catching quality issues early dramatically reduces their impact and resolution cost compared to discovering them later in the process.

Managing Technical Debt

Technical debt accumulates in every product, but not all debt is equally toxic. I've worked on products crippled by technical debt, where even minor enhancements required weeks of refactoring. I've also seen teams paralyzed by perfectionism, refusing to ship valuable features because the implementation wasn't ideally clean. I'm amazed that I still have all the hair on my head and that it's not all gray.

Effective technical debt management distinguishes between strategic debt (deliberately incurred to accelerate time-to-market with a clear repayment plan) and accidental debt (resulting from poor practices without corresponding benefits). It establishes guardrails around maximum acceptable debt levels and creates explicit processes for debt repayment within the regular development workflow.

LinkedIn's engineering organization implemented a "20% time" policy where engineers could devote one day per week to addressing technical debt. This investment improved their development velocity over time as systems became more maintainable, demonstrating that debt repayment isn't just a technical concern but a business performance driver. Their approach balanced immediate feature delivery with long-term sustainability.

Deployment Planning

The gap between engineering "done" and customer value realization contains many pitfalls. I've seen technically excellent products fail in the marketplace because sales teams weren't equipped to sell them, support teams weren't prepared to troubleshoot issues, and customers weren't properly onboarded to new capabilities.

Comprehensive deployment planning addresses go-to-market readiness (sales enablement, marketing campaigns), operational readiness (support training, monitoring capabilities), and customer readiness (communication plans, migration tools). It transforms product shipment from a technical milestone to a coordinated business initiative.

Salesforce's release management process exemplifies thorough deployment planning. Their three-releases-per-year cadence includes extensive preview programs, administrator preparation resources, and tiered rollout strategies. This approach recognizes that successful deployment extends far beyond code promotion to production environments; it encompasses the entire ecosystem needed to deliver customer value.

Operational Scalability

Growth reveals weaknesses in product architecture and operations. I've experienced the pain of rapid user growth exposing scalability limitations, resulting in service degradation exactly when market momentum was strongest. These moments taught me that operational scalability isn't a technical afterthought; it's a core product requirement for success.

Effective scalability planning addresses technical architecture (will the systems handle 10x volume?), operational processes (will support and service delivery models scale?), and organizational structure (will decision-making remain effective as the team grows?). It builds headroom in all these dimensions before growth creates crisis conditions.

Netflix's shift from physical infrastructure to AWS cloud services illustrates strategic scalability planning. They recognized that their growth trajectory would eventually exceed their ability to provision physical servers efficiently. Rather than waiting for capacity constraints to create crises, they proactively redesigned their architecture for cloud deployment, enabling them to scale rapidly as streaming adoption accelerated. Their foresight in addressing scalability before it became an emergency allowed them to capture market share during a critical growth period without operational limitations.

When I owned TangoWire, we grew at such an enormous rate that administration of the site was a challenge. The customer service requests were constant. Uploaded personal images needed examining (there are weird people in this world), and servers needed regular maintenance. We remedied of of these issues with a "control center" application, whereby we could quickly respond to customer requests, automate image cropping, and bring down servers with the untick of a checkbox when they acted up or needed upgrading.

Conclusion

Product leadership ultimately isn't about building products; it's about creating value — for customers, for the business, and for the world. This value creation happens at the intersection of customer needs, business strategy, and operational excellence. Hopefully, the thoughts and examples explored in this chapter won't just help you build better products; they'll help you build products that matter, products that last, and products that contribute meaningfully to your organization's success.

The most valuable product leaders I've worked with aren't those with the most technical knowledge or the most compelling product visions. They're the ones who effectively integrate across domains by connecting customer insights to business outcomes, translating executive priorities into development plans, and balancing short-term delivery with long-term sustainability. They're the bridges that span organizational divides, the translators between technical and business languages, and the integrators who create coherence from complexity.

As you apply these examples to your own product leadership journey, remember that alignment isn't a destination; it's a continuous process of calibration as markets evolve, businesses transform, and customer needs change. The greatest product leaders maintain this alignment not through dogmatic processes but through relentless curiosity, genuine empathy for all stakeholders, and unwavering focus on the fundamental question at the heart of product management: Are we creating sustainable value?

Just a quick note. Reading this chapter alone can be a lot to take in, so don't try to absorb everything at once. Keep coming back to it (and other chapters) regularly. Soon, all these ideas will become second nature. A good idea in this book will become a simple thought or realization in your head — automatically making itself available when you're in an executive meeting someday.

Also, some of these concepts might not be immediately relevant to you. Take some time to explore your current product, revisit this chapter, and I'm sure you'll find some valuable insights that you can apply. Soon it will be like touch-typing — once a pain to learn, now words and even entire sentences just come out automatically and faster than you can think.

Bonus Tip

There's one tip I share with all my coaching clients. It's a label affixed to my laptop, just above the keyboard, and with the following words:

> Valuable / Usable / Possible / Affordable / Maintainable / Durable / Secure / Legal / Ethical

When you start a new product or feature, ask yourself the following questions. If you get hung up with a "no" on one of them, try to work through it, but don't bypass it. If a few of them are raising red flags, turn around and get out!

Valuable: Is this truly valuable to my users? Is there good product-market fit? Does it make sense to build?

Usable: Will my users know how to easily use this feature? Is it intuitive? Will it require education or a support?

Possible: Are we able to build it with our current knowledge and staff?

Affordable: Can we afford to build it? Will it have a respectable ROI or strategic benefit? What are potential cost overruns?

Maintainable: Are we able to maintain it during its entire lifecycle?

Durable: Will it be obsolete quickly? Can it survive all the new AI technologies coming out? Is it built for future growth?

Secure: Is it very secure and protected against hacks?

Legal: Is it fully legal or does it require attorney endorsement?

Ethical: Does it take a privacy-first stance? Is it right? Should we really be building it?

Concepts in Action: PixelRod Software

PixelRod began as a simple photo editing app for designers. Founded by former UI designer Stuart Lewis, the company initially offered basic color correction and resizing tools. However, Stuart understood that strategic vision was the foundation of successful product development.

Unlike competitors who tracked dozens of metrics, PixelRod established a clear North Star: "design time saved." This single metric guided every decision, from feature prioritization to interface design. When considering a new filter library, the team evaluated it against this metric: would it genuinely save designers time?

Their company mission was equally focused: "Empowering creativity through intuitive design tools." This clear mission helped PixelRod avoid distractions. When investors suggested expanding into video editing — a lucrative but different market — the team evaluated this against their mission and declined, choosing instead to strengthen their position in their core market.

Value Proposition and Competitive Positioning

PixelRod's value proposition wasn't just about photo editing; it promised "professional-quality design results in half the time." This specific value proposition guided their development priorities and marketing strategy.

Rather than competing with Adobe as "Photoshop but simpler," PixelRod positioned itself differently. They focused on UI/UX designers who needed quick mockup capabilities. This positioning allowed them to capture a significant share of this specific segment within 18 months, compared to their previously low penetration in the general market.

Solutions for Real Customer Problems

PixelRod exemplifies problem verification before solution development. When their analytics showed designers frequently toggling between their app and design systems, they didn't immediately build an integration feature. Instead, they conducted user interviews and discovered designers were trying to ensure color consistency with their company's brand guidelines.

The solution wasn't a complex integration but a simple color palette import feature that allowed designers to maintain brand consistency. This focus on the actual problem rather than the assumed solution resulted in a feature with 78% adoption within the first month.

Market Timing Analysis

PixelRod's timing was deliberate. They launched when several factors converged:

- The proliferation of design systems in enterprise companies

- Increasing emphasis on brand consistency across digital products
- Growing need for rapid prototyping in agile environments

This timing analysis wasn't accidental. Stuart's team had tracked these trends for years, waiting for the moment when market readiness, technological enablers, and customer mindset aligned.

Customer Journey Mapping

The product team created detailed journey maps showing how designers moved from concept to deliverable. They identified a critical pain point: designers wasted significant time searching for the right assets. This insight led to PixelRod's intelligent asset management system, which automatically organized resources based on project context.

Their journey mapping went beyond functional steps to understand emotional states. They recognized that designers approached revisions with frustration and deadline anxiety. This led to PixelRod's "revision mode" which simplified the process and reduced stress.

Forecasting Revenue and ROI

When pitching to investors, PixelRod didn't just showcase features; they demonstrated financial impact. They projected that their collaborative editing feature would increase team subscription upgrades by 15%, translating to $2.3 million in additional annual recurring revenue.

For resource allocation, they adopted portfolio thinking. They categorized investments as:

- Core investments (improving existing features)
- Growth bets (new capabilities with evidence of demand)
- Breakthrough options (exploratory initiatives with high potential)

They allocated 60% of resources to core, 30% to growth, and 10% to breakthrough initiatives, ensuring balanced investment across their product portfolio.

Pricing Strategy

PixelRod's pricing strategy aligned perfectly with their value delivery. Rather than charging per feature, they created a tiered structure based on time-saving capability:

- Free: Basic editing with standard templates
- Professional: Advanced editing with time-saving presets

- Team: Collaborative features with shared asset libraries

This structure matched how different segments valued the product and created natural upgrade triggers as users' needs evolved.

Cross-Functional Collaboration

PixelRod implemented what they called "product trios." They were small teams consisting of product management, engineering, and design. These teams stayed together through multiple releases, building shared context and understanding.

When launching their collaboration feature, the product team involved sales, marketing, support, and customers from day one. Sales provided input on positioning, marketing prepared educational content, and support trained on troubleshooting. This comprehensive approach ensured all organizational components were aligned before launch.

Resource Capacity Planning and OKRs

PixelRod transformed their planning after a quarter where they missed deadlines on three consecutive releases. Stuart recognized they consistently over-planned by 40%, creating constant disappointment despite extraordinary efforts.

The team implemented "reality-based planning" by analyzing historical data. They discovered maintenance consumed 23% of capacity, unexpected issues took 12%, and onboarding required 8% of senior developer time. They now planned new features around remaining capacity.

During quarterly planning, Stuart began with actual capacity: "We have 420 developer days this quarter. After maintenance, debt reduction, and buffers, we have 215 days for new features. What's the most valuable way to use them?"

This approach transformed prioritization from wishful thinking to strategic trade-offs. When marketing requested a redesigned asset library, Stuart could show exactly what would need to be deprioritized.

PixelRod also implemented a nested OKR method connecting every team's work to company objectives. This gave team members clarity on how their daily work contributed to company success, allowing them to evaluate options against key results rather than subjective preferences.

The combination of realistic capacity planning and aligned objectives created what Stuart called "sustainable momentum," which means consistently delivering high-impact features without team burnout or quality compromises.

Development Velocity and Quality

PixelRod measured outcomes rather than activities. Instead of tracking story points, they measured cycle time from concept to customer availability. This focus on outcomes prevented the illusion of progress without actual delivery.

For quality assurance, they distinguished between different quality dimensions based on feature context. For their collaboration tools, reliability and security received highest priority, while for creative filters, performance and usability took precedence.

Technical Debt Management

PixelRod distinguished between strategic and accidental debt. When entering the enterprise market, they deliberately incurred strategic debt to launch quickly, with a clear repayment plan. They allocated 15% of sprint capacity to debt reduction and measured the impact of this investment on development velocity.

Operational Scalability

As PixelRod grew from 10,000 to 250,000 users in six months, their foresight in scalability planning paid off. They had built headroom in their technical architecture, customer support processes, and team structure before growth accelerated.

They implemented an automated customer support system with escalation paths, allowing them to maintain response times even as volume increased tenfold. This operational scalability enabled them to capture market share during a critical growth period without service degradation.

The PixelRod Checklist

For every new feature, PixelRod's product team asks:

- **Valuable**: Does this genuinely save designers time?
- **Usable**: Can users intuitively understand how to use it?
- **Possible**: Can our team build it effectively?
- **Affordable**: Does the ROI justify the investment?
- **Maintainable**: Can we sustain this feature long-term?
- **Durable**: Will this remain relevant as technology evolves?
- **Secure**: Does this protect user data appropriately?
- **Legal**: Have we addressed all compliance requirements?
- **Ethical**: Are we respecting user privacy and creating positive impact?

This structured approach to product alignment helped PixelRod grow from a small startup to a design industry standard in just three years, with profitability exceeding industry averages by 40%.

Execution Essentials

Strategic Vision

- Establish a clear North Star metric that guides all product decisions instead of tracking too many metrics at once.

- Ensure your product directly serves your company's mission to maintain focus and avoid distractions.

- Create a clear value proposition that translates features into customer benefits and addresses specific pain points.

- Focus on being different in ways that matter to specific customer segments rather than just being "better" than competitors.

- Balance specificity and flexibility in roadmaps. Be precise about near-term deliverables while maintaining strategic flexibility for the future.

Market & Customer Alignment

- Verify problems exist before building solutions by documenting evidence from support tickets, customer interviews, and usage data.

- Analyze market timing by considering technological enablers, regulatory environment, customer mindset, and competitive landscape.

- Map customer journeys to understand not just functional tasks but also emotional, situational, and environmental contexts of product usage.

- Segment users precisely based on behavioral patterns and specific needs rather than trying to build for "everyone."

Business Performance Metrics

- Connect product outcomes to business outcomes by translating improvements in conversion rates or churn into financial impact.

- Consider all costs in your analysis including development hours, ongoing maintenance, technical debt, and opportunity costs.

- Define clear success metrics before development begins and track both resource consumption and value creation.

- Create pricing structures that align with how value is delivered and perceived by different customer segments.

Organizational Synergy

- Enable cross-functional collaboration through structural enablement, cultural reinforcement, and leadership modeling.

- Base capacity planning on honest assessment of true capacity, accounting for maintenance work, technical debt, and unexpected issues.

- Create cascading OKRs that connect company-level goals through department objectives to product initiatives.

- Manage stakeholder expectations through mapping key stakeholders, regular calibration, and proactive adjustment when conditions change.

- Categorize features clearly (core, growth bets, maintenance mode) and develop deliberate strategies for resource allocation.

Execution & Operation

- Focus on outcome metrics like features delivered rather than activity metrics like story points completed.

- Address go-to-market readiness, operational readiness, and customer readiness in deployment planning.

- Build headroom in technical architecture, operational processes, and organizational structure before growth creates crisis conditions.

The Product
Chapter 11

Competitive Analysis

I've learned that true competitive intelligence is both an art and a science — one that separates good product leaders from great ones.

In this chapter, I'll share the approaches I've developed over decades in the trenches. You won't find theoretical models that look pretty in slide decks but fall apart in practice. Instead, I'll give you battle-tested strategies that have helped my teams outmaneuver competitors in markets ranging from enterprise software to consumer apps.

Remember, if your product is already out in the wild, your competitors are running their own analysis on your product right now. The question is: will your strategic response be reactive or proactive? In this chapter, I'm providing you the tools to ensure it's the latter.

Just a heads up, some parts of this chapter might overlap with other chapters and concepts I've covered in this book. I know, I know, it can be a bit confusing. But don't skip over them! I'll be revisiting these topics multiple times, and I'll be doing so from a competitive perspective. This way, you'll get a fresh look at them and see them from a different angle.

Market Positioning & Differentiation

Market positioning isn't something you do once and forget. It's a continuous process of understanding where you fit in the competitive landscape and how to leverage that position for growth.

The most common mistake I see product leaders make is trying to be everything to everyone. A restaurant that serves Italian, Chinese, and Mexican food probably doesn't do any of them particularly well. Your product positioning works the same way. When you try to occupy every position in the market, you end up owning none of them.

Let's say you're a CRM company going head-to-head with Salesforce. You'll start by trying to match their feature set point-for-point. After an exhausting year and declining market share, you'll finally admit you cannot outdo Salesforce at being Salesforce. Instead, you might narrow your focus to small businesses in the service industry and build specialized features that address their unique challenges. Salesforce cannot be everything to everyone, either. With your focus on a specific industry's unique issues, you set yourself up for better success.

The shift works because you stop playing someone else's game and start playing one you can win. That's the essence of effective positioning; finding the intersection of what you do uniquely well and what a specific segment of the market values highly.

Positioning isn't just about standing out; it's about standing out in ways that matter to customers who matter to your business. And you can't do that without understanding exactly how your product solves customer problems compared to alternatives in the market.

Competitor Value Proposition Mapping

As discussed in the previous chapter, value proposition mapping is the process of comparing how your product solves problems versus how competitors do. It's not about features. It's about outcomes.

When I led a workshop on value proposition mapping, I asked teams a simple question: "If your product disappeared tomorrow, what would your customers lose that they couldn't get elsewhere?" The answers revealed their true differentiation. If your team struggles to answer this question concretely, you likely have a positioning problem.

The most effective way to map value propositions is to start with customer jobs-to-be-done, then analyze how your solution and competing solutions address these jobs. For instance, when one project management tool's team identified that, while all competitors helped teams track tasks, their unique value was in how they visualized dependencies between workstreams. This wasn't just a feature difference; it fundamentally changed how program managers could communicate risks to executives.

Don't just take your team's word for this analysis. Talk to customers who evaluated your competitors before choosing you, and vice versa. Their perspective on comparative value is infinitely more insightful than internal speculation. When the company mentioned above did this, they discovered that a feature they thought was a minor convenience was actually the primary reason customers chose them over the market leader.

Remember that value propositions evolve as markets mature. What differentiates you today may become table stakes tomorrow. Netflix's original value proposition was DVD delivery without late fees. Now it's about original content and personalized recommendations. Your mapping should anticipate these shifts, not just document the current state.

Pricing Against Competitors

Pricing isn't just about revenue. It's a positioning tool that signals your value to the market. Understanding where you sit in the price-to-value spectrum is important.

I've made the mistake of underpricing a premium product based on competitive pressure. We thought lower prices would accelerate growth. Instead, it confused customers about our positioning and actually slowed sales because prospects questioned our quality. When we raised prices by 30%, conversion rates improved. This is a counterintuitive but common pattern for products that get their value from quality or exclusivity.

Your pricing analysis should look at direct and indirect competitors. When Apple prices the iPhone, they're not just considering Samsung's pricing. They're considering how consumers allocate their discretionary spending across categories. Similarly, your enterprise software isn't just competing with other software; it's competing with consultants, outsourcing, and the option to build in-house.

Analyze not just price points but pricing models. Subscription, perpetual licensing, usage-based, and freemium all create different customer behaviors and competitive dynamics. Slack's freemium model fundamentally changed how collaboration tools compete by creating viral bottom-up adoption that circumvented traditional enterprise sales processes.

Most importantly, pricing strategy must align with your overall differentiation. Walmart's "everyday low prices" works because their entire supply chain is optimized for cost efficiency. If your product is differentiated on quality or unique functionality, competing primarily on price is usually a mistake. At one company, they commanded a much higher price premium by focusing their messaging on the specific metrics where their solution outperformed competitors by the largest margin.

Comparative Brand Perception

Brand perception is how customers view your brand versus alternatives. It's subjective but measurable, and often more influential than feature differences.

I've seen startups with objectively better technology struggle against established competitors with stronger brands. Trust is the currency of business, and brand is the primary vehicle for communicating trustworthiness. This is why Apple can enter new categories like watches or credit cards with immediate credibility, while startups with similar offerings must fight for every early customer.

The most effective way to assess comparative brand perception is through qualitative research. Listening to how prospects and customers talk about you versus competitors reveals insights no survey can capture. Conduct quarterly interviews where you simply ask people to describe your brand and your top three competitors in their own words. The patterns that emerge are invaluable.

Quantitative measures matter too. Net Promoter Score (NPS) comparisons, sentiment analysis of social mentions, and share of voice metrics all provide data points for your assessment. But don't get lost in vanity metrics. Ultimately, brand perception matters mostly when it influences purchase decisions and customer loyalty.

Remember that different segments may have radically different brand perceptions. Enterprise buyers may view Microsoft as reliable and secure, while developers might see them as cumbersome and outdated. Your assessment should account for these variations across your target segments.

Don't get confused between brand and identity. Your company's logo is a badge which allows your customers to recognize your brand and it's part of your identity. Your brand encompasses your reputation, quality and fit of your products, and level of trust you project in business. When you work with a graphics artist to develop an identity package, you're asking them to create your logo, letterhead, and various forms of bling. You work with marketing and public relations people to develop your brand.

Feature-to-Feature Benchmarking

While features aren't the whole story, direct comparison of capabilities that matter most to users is still an essential part of competitive analysis.

The key word here is "matter." I've seen too many product teams create exhaustive feature comparison spreadsheets where 80% of the items have minimal impact on purchase decisions. Focus your benchmarking on the capabilities that directly address your customers' most important jobs-to-be-done.

When I was once developing a new analytics platform, we initially tracked 78 feature comparisons across competitors. After customer interviews, we narrowed this to 12 capabilities that truly influenced buying decisions. This simplified not just our analysis but our entire product strategy. We stopped spreading resources across some features and doubled down on the ones that really moved the needle.

Don't limit your benchmarking to binary yes/no comparisons. Assess implementation quality and ease of use. A competitor might technically have a feature, but if it requires professional services to configure or has poor usability, that's a meaningful difference. Zoom didn't win by inventing video conferencing; they won by making it work more reliably and with fewer clicks than alternatives.

Finally, use benchmarking to identify patterns, not just gaps. Is a competitor consistently ahead in a particular capability area? Is their UI super simple, yet more stunning than others? Are they systematically addressing a customer segment you're overlooking? These patterns often reveal strategic priorities more clearly than individual feature differences.

Example: TrekStuff

TrekStuff started as yet another travel gear company in a crowded marketplace. Their initial strategy was trying to compete with established brands like REI and Patagonia across all outdoor categories — backpacks, tents, clothing, you name it.

After a frustrating year of mediocre sales and spreading their resources too thin, CEO Shana Hayes had an epiphany during a team meeting. "We can't outdo REI at being REI," she admitted. "We're trying to be everything to everyone and ending up being nothing special to anyone."

The team refocused entirely on ultralight backpacks specifically designed for digital nomads — people working remotely while traveling. Instead of competing on features that bigger companies could easily match, they positioned their products around a specific outcome: helping remote workers carry tech gear comfortably while maintaining a professional appearance.

Their signature product became the "OfficeNomad" backpack that didn't look out of place in a coffee shop meeting but could handle a weekend trek. While competitors' bags were either too rugged-looking for professional settings or too stylish for outdoor use, TrekStuff occupied the intersection of professional and adventurous.

When they interviewed customers who chose them over bigger brands, they discovered something surprising — what sealed the deal wasn't their water-resistant materials (which they thought was their edge), but their unique cable management system that kept chargers and cords organized. This insight drove their next generation of products.

By targeting a specific customer segment with particular needs, TrekStuff stopped playing someone else's game and created one they could win.

Competitive Intelligence Gathering

Competitive intelligence isn't about spying. It's about collecting and analyzing publicly available information, so you can make better strategic decisions.

When done right, competitive intelligence feels like having a superpower. You can anticipate competitors' moves, identify market shifts before they become obvious, and make more confident product decisions. When done wrong, it's an expensive distraction that creates fear-based reactivity rather than strategic clarity.

Win/Loss Analysis

Win/loss analysis examines why customers choose or reject your product over competitors. It's the most direct feedback on your competitive position.

One team assumed they knew why they won or lost deals. Sales would tell them "they went with Competitor X because of price" or "we won because our UX is better." When they started doing structured win/loss interviews, they discovered these explanations were often dramatically oversimplified or entirely wrong. In one eye-opening case, what sales reported as a pricing loss was actually because their product lacked a specific security certification the customer deemed non-negotiable.

The key to effective win/loss analysis is talking directly to the decision-makers, not just relying on sales team reports. At minimum, aim to interview 20% of your wins and losses, focusing on deals that represent your target customer segments. Use a consistent interview guide, but leave room to explore unexpected insights.

Document patterns, not just individual data points. If you lost three enterprise deals because of compliance, that's a pattern worth addressing. If you lost one deal because of a highly specific feature request, that might just be noise. The art is distinguishing signal from noise without dismissing potentially important feedback.

Most importantly, create a closed feedback loop between win/loss findings and product strategy. Review win/loss patterns monthly with the product team and quarterly with executives. This direct line of insight has influenced roadmap priorities more than any other single input. Because it's such an esoteric topic to meet over, incorporate it in existing monthly or quarterly meetings.

Applying SWOT

The SWOT method (Strengths, Weaknesses, Opportunities, Threats) may seem basic, but it's endured for a reason. When it's applied rigorously, it works.

The mistake most teams make with SWOT is treating it as a one-time exercise rather than an evolving analysis. Markets change, competitors adapt, and your own capabilities develop. Your SWOT analysis should be a living document, updated quarterly at minimum.

When I conduct SWOT analyses, I insist on specificity and evidence. "Great UX" isn't a strength unless you can demonstrate how and why your UX outperforms competitors in ways that matter to customers. "Market consolidation" isn't a threat unless you can articulate how specific consolidation scenarios would impact your business.

The most valuable insights often come from the intersections between quadrants. How can you use your strengths to capitalize on opportunities? How might competitors exploit your weaknesses? These dynamic relationships reveal strategic options that static feature comparisons miss.

I've found it valuable to create separate SWOT analyses for different market segments or customer personas. Your strengths relative to competitors often vary dramatically across different use cases or customer types. A one-size-fits-all SWOT analysis masks these important nuances.

Strengths	Weaknesses
• **Superior UI** • **Micro front end modularity for faster expansion** • **AI powered recommendations** • **Privacy-first Policies**	• **Only in English (no localization)** • **UI issues with some browsers** • **Complex product filtering** • **Only North America**
Opportunities	**Threats**
• **Brand collaborations** • **European expansion** • **New personalization engine** • **Influencer partner program** • **Affiliate program**	• **Competitors using same React UI framework** • **AI moving too fast to keep up** • **Trade tariffs too high — trade war heating up** • **No exclusive rights to best products**

An example of a SWOT created for a new e-commerce website.

Digital Footprint Analysis

Digital footprint analysis involves studying competitors' online presence, messaging, and engagement to infer strategic priorities and market positioning.

This isn't about vanity metrics like social followers or website traffic. It's about extracting strategic signals from the digital choices competitors make. What keywords are they bidding on? How has their messaging evolved over time? Which customer stories do they highlight? What roles are they hiring for?

At a friend's company, they noticed a competitor suddenly increasing their ad spend on keywords related to a vertical market space they dominated. Simultaneously, the the competitor published three case studies from that industry and hired two sales specialists with vertical expertise. These signals, combined with rumors from the sales team, gave them 4-6 months of advance notice that their competitor was planning a major push into the same territory. My friend used that time to strengthen their relationships with key accounts and accelerate their roadmap for that vertical.

The tools for this analysis range from free (Google Alerts, social listening) to enterprise (competitive intelligence platforms). But the tools matter less than the analytical methods you apply to the information. Look for changes over time, discrepancies between messaging and actual offerings, and patterns across different channels.

Remember that competitors' digital footprints reveal not just current positioning but future direction. Job postings, partner announcements, and leadership changes often telegraph strategic shifts months before product launches or official pivots.

When working with a product which has competition, I set aside a full day or two every couple months and to comb the web for my competitors' digital footprints. I look at their press releases, any news articles on Google, customer reviews in marketplaces, and their search engine SEO and keyword usage. Don't think for a second that they're not doing the same research.

Customer Review Mining

Customer review mining involves extracting patterns from what users say about your competition across review sites, social media, and communities.

This is one of the most underutilized competitive intelligence sources I've seen. Your competitors' customers are publicly sharing detailed feedback about product limitations, support issues, and unmet needs — invaluable intelligence that costs nothing but time to collect.

At an LED lighting company, we were entering a crowded market and we analyzed over 1,000 reviews of the top five competitors. We found a consistent pattern of complaints about product longevity and light quality. These were pain points none of the competitors had effectively addressed. This insight directly informed our product differentiation strategy and messaging, helping us carve out a viable position despite entering a flooded market. I was then the first product lead to deploy a much cooler and more color accurate "remote phosphor" technology. It was a huge hit, until the Chinese closely monitored our digital footprint and knocked off that same product a few years later.

The key is looking beyond star ratings to qualitative patterns. What specific use cases generate the most positive reviews? What limitations do users mention repeatedly? How do sentiment patterns differ across customer segments or use cases? Tools like sentiment analysis can help scale this process, but human judgment remains essential for extracting strategic insights.

Don't limit your analysis to dedicated review platforms. Communities like Reddit and industry forums often contain more candid, detailed, and professional discussions of product experiences than formal reviews. These sources can reveal emerging pain points or use cases before they appear in official channels.

Strategic Response Planning

Competitive intelligence without strategic response is just expensive trivia. The true test of your competitive analysis is whether it drives better decisions.

I've seen teams spend thousands of hours on competitive research that sits in shared drives and informs nothing. Conversely, I've seen lightweight but focused competitive intelligence programs fundamentally shape product strategy and go-to-market execution.

The difference is in how systematically you translate competitive insights into action. Here's how to ensure your analysis drives results rather than collects dust.

Gap Identification

Gap identification means finding market needs which competitors aren't addressing. It's the white space where you can win without engaging in head-to-head competition.

The most valuable gaps aren't always feature gaps. Sometimes they're experience gaps, support gaps, or business model gaps. Zoom didn't win by offering video features WebEx lacked; they won by eliminating the friction in starting and joining meetings. The gap they exploited was experiential, not functional.

One company entered a mature market dominated by two enterprise players. Rather than matching their extensive feature sets, the company found that mid-market customers were drastically underserved; forced to buy complex enterprise solutions with long implementation cycles and high costs. They then developed a streamlined product specifically for this segment, focusing on fast implementation and core workflows while deliberately omitting the complexity that bogged down their competitors. Within 18 months, they owned nearly a quarter of the mid-market segment.

To identify meaningful gaps, combine multiple inputs: customer interviews about competitor limitations, win/loss patterns, review mining, and sales team feedback. Look for consistent patterns across these sources rather than chasing every potential gap.

Remember that not all gaps are worth filling. Evaluate each potential opportunity against your strategic objectives, core capabilities, and resource constraints. Sometimes the right move is to leave certain gaps for others while you focus on areas where you have sustainable advantages.

Roadmap Prioritization

Roadmap prioritization means aligning development priorities based on the competitive landscape while staying true to your strategic vision.

The tension between being customer-driven, vision-driven, and competition-driven is something every product leader faces. I've seen teams swing to both extremes; either ignoring competitive realities or frantically chasing every move by their competitors. Neither works.

At a large bank and for an online banking redesign, I used a simple but effective method: the 70/20/10 rule for my roadmap allocation. Approximately 70% of our resources would go to customer needs and strategic vision, 20% to competitive response or parity on must-have capabilities, and 10% to more experimental, leapfrog initiatives. This balanced approach kept my proposal focused on our differentiation while ensuring we didn't fall behind on table stakes capabilities.

When competitive inputs influence your roadmap, be transparent about the reasoning with your team. "We're building this because Competitor X has it" is poor motivation. "We're building this because it's becoming an expected capability in our category, as evidenced by Competitor X's traction with this feature" frames the decision in terms of meeting evolving market needs.

Remember that the goal isn't feature parity across the board. Strategic feature gaps — areas where you deliberately choose not to compete — can be as important as your areas of differentiation. Clarity about what you won't build is often more strategically valuable than an ever-expanding list of what you will build.

Example: SoundBrat

SoundBrat, a collaborative music software suite, was getting crushed trying to match feature-for-feature with the big players. Their small team was exhausted, and their product had become a confusing mess of half-implemented features. Then their new product lead, Alex, did something simple but effective.

Alex interviewed 30 indie musicians who'd tried SoundBrat and competitors. A clear pattern emerged — while the big platforms excelled for studio recording, remote collaboration was painful. Musicians were cobbling together file sharing, video calls, and clunky workarounds just to jam together online.

This was their gap! Instead of trying to be everything to everyone, SoundBrat refocused entirely on real-time collaboration. They applied the 70/20/10 roadmap rule: 70% went to building their killer real-time collaboration features, 20% to maintaining just enough recording functionality to be viable, and 10% to experimental features like AI-assisted mixing.

Within eight months, SoundBrat became the go-to platform for distributed music teams. They weren't trying to compete with ProTools anymore — they were solving a problem ProTools wasn't addressing at all. Their user base tripled, and their NPS scores jumped from the 40s to the 70s.

The lesson? Sometimes winning means playing a different game entirely, and ruthlessly prioritizing the features that serve your unique position in the market.

Preemptive Innovation Strategy

Preemptive innovation strategy means staying ahead of predictable competitive moves through forward-looking development rather than reactive responses.

The best competitive response isn't responding at all. Instead, it's making your competitors respond to you by anticipating and leapfrogging their next moves. This requires a lot of understanding of your competitors' patterns, constraints, and strategies.

Again, while working for a large bank who has mortgage assets over $1 trillion, I proposed online banking leapfrog features; some to compete with TrueBill (now the Rocket Money app). Many of our customers heavily relied on account information aggregators like TrueBill, even more than their own bank's app. However, our newly recruited executives didn't see it this way, but then I don't believe they understood the market space, either.

The distant competitor's app team kept preemptively building features and advantages for our own customers. They were eventually purchased and rebranded by Rocket Mortgage as the Rocket Money app, now an arm of the now largest mortgage originator in the country.

A small, innovative app can strategically enter your bedrock business if you're not careful. For instance, access to an individual's banking information is powerful. It can reveal who's a good risk, who has funds for a new home, how much they can afford, how stable their income is and from where it originates, and so much more. A distant, strategic competitor came in a perfectly legal back door and obtained our own data!

To develop preemptive strategies, study competitors' innovation patterns and decision-making speed. Some organizations consistently lead in certain areas while following in others. Some are quick to announce but slow to deliver. Some reliably pursue certain types of acquisitions (wink, wink). These patterns let you predict their likely next moves with surprising accuracy — if you're a good product leader, of course.

Continuous Monitoring & Analysis

Competitive analysis isn't a project; it's a process. Markets evolve, competitors adapt, and one-time analyses quickly become obsolete.

I learned of one well known software company that did annual competitive deep dives. By the time the research was synthesized and shared, much of it was already outdated. This gave us a dangerous false confidence. They thought they understood our competitive position because we had recently completed an analysis, but the market had already shifted.

Building sustainable competitive intelligence capabilities requires systems, not just periodic efforts. Here's how to establish ongoing monitoring that keeps your understanding current without creating overwhelming noise.

Market Share Tracking

Market share tracking means measuring your slice of the pie over time, using both quantitative metrics and qualitative indicators.

In mature markets with public companies, share tracking might use reported revenues or unit sales. In emerging categories or private company ecosystems, you'll need proxy metrics: web traffic comparisons, social share of voice, app download estimates, or third-party surveys.

At a SaaS company I built and led, I created a quarterly market share estimate using a combination of public financial data for listed competitors, growth patterns inferred from hiring trends and funding announcements for private competitors, and our own customer count and revenue figures. While not perfectly precise, this gave us a consistent way to track directional shifts in relative position.

Don't just track overall share. Segment your analysis by customer size, industry, and geography. We once discovered we were gaining share overall while losing ground in a strategic segment, a crucial insight that prompted a targeted feature and partner initiative to address the gap.

Remember that market share is a lagging indicator. By the time it shifts significantly, the causes are often months or years in the past. That's why market share tracking must be combined with earlier indicators of competitive position changes.

Feature Release Monitoring

Feature release monitoring means staying current on competitors' new capabilities to understand their strategic direction and potential threats to your positioning.

This isn't about paranoia or feature envy; it's about maintaining situational awareness. At minimum, assign team members to follow competitor release notes, announcements, and product blogs. More sophisticated approaches include using competitive intelligence platforms or establishing alerts for changes to competitors' websites, help centers, and API documentation.

Maintain a "competitor release radar"; a simple monthly summary of significant launches across your competitive set. This will keep your team informed without creating distraction. Each summary might include three sections: releases relevant to your strategic differentiation, releases that might influence your roadmap priorities, and releases that are interesting but require no action.

The most valuable insights often come not from individual feature releases but from patterns across multiple releases. Does a competitor consistently invest in a particular area? Are they gradually assembling capabilities that could enable a pivot into an adjacent market? These patterns reveal strategic intent more clearly than press releases or marketing messages.

Remember that feature launches don't always reflect actual usage or traction. A competitor might announce a capability with fanfare, but if customers don't adopt it, the competitive impact is minimal. When possible, gather intelligence on feature adoption and user sentiment, not just release announcements.

Pricing Change Alerts

Pricing change alerts mean detecting shifts in competitor monetization strategies that could impact your market position or unit economics.

Price changes signal more than just tactical adjustments. They very often reflect fundamental strategic repositioning or response to market pressures. A competitor dropping prices might indicate commoditization pressures, while new pricing tiers or models might signal expansion into new segments.

To keep an eye on pricing changes, use a mix of automated monitoring (website scraping for public pricing pages) and manual tracking (sales team reports of competitive deals). When you notice a competitor offering a lower-tier product, quickly check if it's missing important features. This could be a sign that they're trying to cut costs while still keeping their main products.

Don't just track list prices. Monitor discounting patterns, contract term changes, bundling strategies, and promotional offers. A competitor maintaining list prices while increasing discounting flexibility is effectively cutting prices without the market visibility of a formal price reduction.

When you detect significant pricing changes, don't rush to match them. First, understand the strategic intent and likely impact. Is this a temporary promotion or permanent shift? Is it targeted at a specific segment or across-the-board? Your response should address the strategic implications, not just the price point itself.

Competitive Landscape Shifts

Competitive landscape shifts involve identifying new entrants and evolving threats that could fundamentally change market dynamics.

The most dangerous competitors are often not your current direct rivals but adjacent players expanding into your space or new entrants with disruptive models. For example; taxi companies in NYC were worried about new taxi services and newly issued tokens (transferrable taxi permit medallions). They were then blindsided by the ridesharing company Uber.

Maintain a "competitive perimeter watch"; monitoring direct competitors and companies in adjacent spaces who could potentially enter your market. Track their product expansions, strategic hires, and partnership announcements for early warning signs.

Don't forget about potential consolidation. When two smaller competitors merge, or a larger player acquires a specialized solution, the competitive dynamics can change overnight. Maintain scenarios for the most likely or impactful potential combinations and consider how they would affect your strategy.

Finally, track shifts in the funding landscape. Changes in where investors are placing bets often precede market shifts by months or years. When you notice a sudden increase in funding for vertical-specific solutions in your space, it may signal the beginning of market fragmentation that will influence your segmentation strategy.

Conclusion

Remember, competitive analysis isn't about obsessing over competitors. It's more about developing a clear-eyed view of market realities so you can make better strategic choices. The goal isn't to react to every move by your competitors, but to understand the landscape deeply enough to make proactive decisions that create sustainable advantage.

The best product leaders I know share a common trait: they're neither afraid of competition nor incensed by it. They maintain respectful awareness while staying focused on their product's value and their vision for its future.

Concepts in Action: BrewChoice

BrewChoice started as a small subscription service delivering craft beer kits to home brewers. Founded by three friends with a passion for brewing, the company initially focused on delivering high-quality ingredients with fool-proof instructions. After two years of steady growth, they faced increasing competition from both established brewing supply companies and new subscription startups entering the market.

Market Positioning & Differentiation

When larger competitor BrewSupreme entered the market with nearly identical kits, BrewChoice initially panicked. They scrambled to match BrewSupreme's extensive catalog, adding wine kits and kombucha options. Sales plateaued despite the expanded offerings.

"We were trying to be everything to everyone," recalls CEO Benny Bartlett. "We were the restaurant serving Italian, Chinese, and Mexican food — and doing none of them particularly well."

The turning point came when BrewChoice narrowed their focus to specialized small-batch seasonal recipes with locally sourced ingredients. They stopped competing directly with BrewSupreme's broad catalog and instead created limited-edition collaborations with microbreweries. This positioning shift allowed them to charge premium prices while establishing a unique identity in the market.

Competitor Value Proposition Mapping

BrewChoice conducted a workshop asking a simple question: "If our product disappeared tomorrow, what would customers lose that they couldn't get elsewhere?"

The answers revealed their true differentiation wasn't just ingredients; it was the educational component. While all competitors helped customers brew beer, BrewChoice uniquely focused on teaching the science behind brewing through interactive online sessions.

Customer interviews confirmed this finding. Many subscribers had evaluated BrewSupreme but chose BrewChoice specifically for the learning experience. What the team thought was a minor feature — their "Brew Science" video series — was actually a primary decision driver.

Pricing Against Competitors

Initially, BrewChoice priced their subscription below market leaders, assuming lower prices would accelerate growth. Instead, it confused customers about their positioning and slowed sales as prospects questioned their quality.

"When we raised our prices by 25%, our conversion rates actually improved," explains CFO Michael Rivera. "We realized we were competing in a premium segment where price signals quality."

Their pricing analysis extended beyond direct competitors. They discovered they weren't just competing with other brewing kits; they were competing with craft brewery tours, cooking classes, and other experiential learning opportunities. This insight led to a pricing model that emphasized the educational value, not just the physical ingredients.

Comparative Brand Perception

Through quarterly customer interviews, BrewChoice identified how customers viewed them versus alternatives. While BrewSupreme was seen as "reliable" and "comprehensive," BrewChoice was described as "educational," "authentic," and "community-focused."

This perception varied across segments. Experienced brewers valued BrewChoice's technical depth, while beginners appreciated their approachable instruction style. Understanding these nuanced perception differences helped tailor marketing messages for different audiences.

Feature-to-Feature Benchmarking

Rather than tracking dozens of features, BrewChoice focused on benchmarking the 12 capabilities that truly influenced buying decisions. They assessed not just whether competitors had similar features but implementation quality and ease of use.

For example, while several competitors offered brewing video tutorials, BrewChoice's were consistently rated higher for production quality and instructional clarity. This qualitative difference became a key selling point in their marketing.

Win/Loss Analysis

BrewChoice implemented structured win/loss interviews with 20% of their wins and losses. These conversations revealed insights that transformed their strategy.

"We assumed we were losing to BrewSupreme because of their wider selection," notes Marketing Director James Tong. "But our interviews revealed we were actually losing because our website lacked mobile optimization for checkout. Experienced brewers browsed recipes on their phones while in their brewing spaces."

This direct insight led to immediate mobile experience improvements, resulting in a 15% conversion rate increase within two months.

Applying SWOT

BrewChoice created separate SWOT analyses for different market segments. For beginners, their strength was simplicity and education. For experienced brewers, it was recipe authenticity and ingredient quality.

By examining the intersections between quadrants, they identified a key opportunity: using their educational strength to address the threat of DIY brewing guides proliferating online. This led to the development of their highly successful "Master Class" series featuring renowned brewers.

Digital Footprint Analysis

The team noticed competitor BrewTastic suddenly increasing search ad spend on keywords related to brewing equipment. Simultaneously, they published three equipment-focused articles and hired a product manager with hardware experience.

These signals gave BrewChoice advance notice that BrewTastic was likely expanding into brewing equipment. BrewChoice used this intelligence to accelerate their own equipment partnership strategy, securing exclusive distribution deals with premium manufacturers before BrewTastic's official launch.

Customer Review Mining

BrewChoice analyzed over 800 reviews of the top three competitors. A consistent pattern emerged: complaints about ingredient freshness and packaging waste. This insight directly informed their product strategy, leading to their innovative vacuum-sealed ingredient pouches and compostable packaging that became a major differentiator.

Gap Identification

Rather than matching BrewSupreme's extensive recipe library, BrewChoice identified an experience gap. Competitors focused on recipe variety but neglected the brewing process itself. BrewChoice developed augmented reality tutorials that walked brewers through each step using their smartphone cameras; solving a problem no competitor had addressed.

Roadmap Prioritization

BrewChoice adopted a 70/20/10 method for their roadmap: 70% focused on customer needs and strategic vision, 20% on competitive parity features, and 10% on experimental initiatives.

"We're transparent about competitive influences on our roadmap," says Product Manager Lisa Creston. "We don't build features just because BrewSupreme has them. We build capabilities that are becoming expected in our category, while staying focused on our educational differentiation."

Preemptive Innovation Strategy

By studying competitors' innovation patterns, BrewChoice predicted Brew-Supreme would eventually launch brewing equipment packages. Rather than waiting to respond, they preemptively partnered with artisan equipment makers to create exclusive co-branded brewing kits, establishing a premium position before BrewSupreme entered the category.

Continuous Monitoring & Analysis

BrewChoice established ongoing competitive monitoring systems:

- A quarterly market share estimate using competitor hiring trends, funding announcements, and their own subscriber growth rates.
- A "competitor release radar" providing monthly summaries of significant launches across competitors.
- Automated pricing change alerts tracking not just list prices but discount patterns and promotional offers.
- A "competitive perimeter watch" monitoring adjacent players who might enter their market.

"The goal isn't to react to every competitor move," emphasizes CEO Bartlett. "It's to understand the landscape deeply enough to make proactive decisions. We maintain respectful awareness while staying focused on our unique value proposition: transforming home brewers into brewing experts."

This balanced approach to competitive analysis helped BrewChoice grow from a niche subscription service to the leading educational brewing platform, with revenue growth outpacing the broader home brewing market by 3-times over five years.

Execution Essentials

Market Positioning & Strategic Differentiation

- Don't try to be everything to everyone. Focus on what you do uniquely well that a specific segment values highly.

- When competing with larger players, avoid playing their game. Find your own game where you can win.

- Ask: "If your product disappeared tomorrow, what would customers lose that they couldn't get elsewhere?"

- Start with customer jobs-to-be-done, then analyze how your solution and competing solutions address these jobs.

- Regularly reassess your positioning as markets evolve to ensure continued differentiation.

Pricing and Brand Perception

- Pricing is a positioning tool that signals your value to the market, not just a revenue lever.

- Consider both direct competitors and indirect alternatives when analyzing pricing strategy.

- Conduct qualitative research by asking prospects to describe your brand and competitors in their own words.

- Don't confuse brand (reputation and trust) with identity (logo and visual elements).

Feature Benchmarking and Gap Analysis

- Focus benchmarking on capabilities that directly address customers' most important jobs-to-be-done.

- Assess implementation quality and ease of use, not just the presence of features.

- Look for experience gaps, support gaps, or business model gaps, not just feature gaps.

- Combine multiple inputs (customer interviews, win/loss patterns, reviews) to identify meaningful gaps.

- Be selective about which gaps to fill, as not all represent worthwhile opportunities.

Competitive Intelligence Gathering

- Talk directly to decision-makers in win/loss analysis, not just relying on sales team reports.

- Update your SWOT analysis quarterly as markets change and competitors adapt.

- Extract patterns from user reviews to identify pain points or unmet needs that competitors haven't addressed.

- Study competitors' messaging evolution, keyword focus, and hiring patterns to infer strategic priorities.

- Include communities like Reddit and industry forums which often contain candid, detailed discussions.

Strategic Planning and Continuous Monitoring

- Balance customer-driven, vision-driven, and competition-driven priorities in your roadmap.

- Study competitors' innovation patterns to predict and leapfrog their likely next moves.

- Treat competitive analysis as an ongoing process, not a one-time project.

- Maintain a "competitor release radar" to track significant launches without creating distraction.

- Track shifts in the funding landscape as a leading indicator of future market directions.

The Product
Chapter 12

Product-Led Growth

As a product manager and director who has weathered more storms than I care to count, I've developed a particular lens through which I view product growth. It's not just about hockey stick metrics or flashy launch parties. True product growth — the kind that builds empires rather than sandcastles — is methodical, strategic, and downright unsexy in its day-to-day execution.

In this article, I'll share what I've learned about creating sustainable product growth; the strategies that have consistently delivered results across different industries and product types. My goal is to provide methods that you can adapt to your specific situation, whether you're leading a B2B SaaS platform, a B2C mobile app, or both.

What follows isn't a collection of growth hacks or quick wins. Instead, I offer a sustainable approach to product growth that builds value over time and creates defensible market positions. Because at the end of the day, we're not just trying to grow a product. We're trying to build something that matters; something sticky to customers.

A banking director once told me that my growth strategy was moving "at the pace of continental drift." It's a wise statement. After all, continents may move slowly, but they're impossible to stop once they gain momentum.

Product-Market Fit

Before diving into growth tactics, we need to address the foundation of all sustainable growth: genuine product-market fit. I've seen too many teams push for growth before they've truly achieved this crucial milestone, only to watch their metrics crash when artificial acquisition tactics run out of steam.

Product-market fit isn't a true or false equation; it exists on a spectrum. At a basic level, it means you've built something people want and are willing to pay for. But then **true** product-market fit goes deeper. It's when it becomes the best solution to a meaningful problem for a specific group of people.

The signals of product-market fit are both quantitative and qualitative. Retention curves that flatten. Expanding usage within existing accounts. Word-of-mouth referrals occurring organically. Sales cycles that shorten over time. Customer feedback that shifts from "nice to have" to "can't live without."

In my experience, the most telling sign is when customers become genuinely upset at the thought of losing access to your product (or one of its features). I recall conducting an A/B test on a small segment of our customers and where we temporarily disabled a feature to measure its impact. Within hours, our support team was flooded with frustrated users. That's when I knew we had something worth scaling.

For one online community product I managed, we spent eight months iterating before pursuing growth. This patience allowed us to increase our subscription rate from 5% to 8% and increase our MRR exponentially. When we finally turned on the growth engines, our revenue model was solid enough to support sustainable expansion.

The lesson? Resist the temptation to grow prematurely. As the old saying goes, there's no point scaling something that doesn't work. You'll just end up with a bigger version of a piece of junk.

I've found that explaining product-market fit to some eager stakeholders is like describing water to a fish; they often can't see what they're swimming in. My favorite metaphor is comparing it to dating: user acquisition is getting someone to agree to a first date, but product-market fit is when they're already planning the next date before the first one ends.

The Four Pillars of Sustainable Growth

Once you've established solid product-market fit, sustainable growth depends on four interconnected pillars: acquisition, activation, retention, and expansion. Let's examine each one of them...

Thoughtful Acquisition

Acquisition is often where teams focus most of their growth efforts, but it's actually the least leveraged pillar. Without strong activation and retention, new users simply create a leaky bucket situation.

That said, thoughtful acquisition is still essential. The key word here is "thoughtful." This means targeting the right users, as not all users are created equal. Focus on acquiring those who match your ideal customer profile, the ones most likely to find value in your product and stick around.

It also means selecting channels with CAC:LTV alignment. Different acquisition channels produce different types of customers. Some channels might bring in users with high lifetime value (LTV), justifying a higher customer acquisition cost (CAC). Others might deliver lower-value users, requiring more efficient acquisition methods.

Most importantly, build sustainable acquisition loops. The best acquisition strategies become more efficient over time, not less. This often involves creating flywheel effects where existing customers help bring in new ones.

I once worked on a partnership program where our initial acquisition strategy relied heavily on search engines and social media. Our acquisition cost was reasonable at first, but it got very expensive as competition increased. Meanwhile, we noticed that partners who came through referrals had nearly double the retention rate of those from paid channels and we were able to offer the referring partners residual commissions.. We pivoted our strategy to focus on "Become a Partner" links within the footer of the site. Over time, the percentage of new partners coming through organic referrals grew from 12% to 47%, dramatically reducing our acquisition costs.

The lesson? Build acquisition channels that improve with scale rather than deteriorate. This often means prioritizing word-of-mouth, network effects, and content that provides value independent of your product.

When explaining acquisition strategy to my teams, I like to use the party analogy: paid acquisition is like hiring people to attend your party, while organic growth is when people are excited to come and bring their friends. Both approaches can work, but only one is sustainable in the long run.

Meaningful Activation

Activation is where potential turns into something concrete. It's the moment when a user experiences the core function and value of your product for the first time. This very action and moment is critical for setting the stage for keeping your customer long-term.

Identifying your product's "Aha! moment" is crucial. What is the key action or experience that correlates with long-term retention? For Dropbox, it might be saving a file; for Twitter, it might be following enough accounts to create an engaging feed.

Streamlining the path to value is equally important. Remove unnecessary friction between signup and the "Aha! moment." This often means delaying feature introductions, minimizing form fields, and providing just enough guidance without overwhelming new users.

Personalizing the onboarding experience can dramatically improve activation rates. Different user segments may have different paths to value. Tailoring the onboarding flow to specific use cases gets users to value faster.

For an online banking product I managed, we discovered through data analysis that new customers who checked their accounts more than once within the first week were much more likely to inquire about other financial services and products. This insight transformed our new customer strategy from inundating them with banner ads to making our accounts and details pages super simple and productive — what I call "greasing the funnel." We made what they came for much easier and streamlined. The result? Our activation rate for other products increased 5% in three months.

Remember that activation isn't just about functional completion of steps. It's also about emotional engagement. Users need to feel successful and see the potential solution your product offers to solve their particular problem.

I often tell my product managers that good activation is like teaching someone to ride a bike; you need to run alongside them, holding onto the seat just long enough for them to feel like they're balancing on their own, then let go at the exact right moment.

Deliberate Retention

Continuously delivering new value ensures even satisfied users stay engaged. They may use another product if they stop discovering new value over time. A thoughtful product roadmap should balance new capabilities with refinement of existing ones.

One particularly effective tactic I deploy is "usage depth" initiative. I found that customers who used at least 5 of our 12 core features had renewal rates above 90%, compared to 40% for those using fewer features. My team created adoption campaigns for underutilized features, resulting in a 24% increase in multi-feature usage and a corresponding lift in renewals.

The best retention strategies make churning feel like more work than staying. This doesn't mean creating artificial lock-in, but rather continuously delivering value that would be painful to lose.

I've found that explaining retention to stakeholders works best through the lens of "relationships." Acquisition is like getting someone's phone number, activation is like a great first date, but retention is building a lasting relationship where both parties continue to grow together.

Ask your customers which features they value most and which features (possibly in competing products) you should be including and which they'd like to have. Retention requires customer outreach. Do it.

Strategic Expansion

Expansion — usually getting your existing customers to pay you more over time — is often the most overlooked growth lever, yet it can be one of the best. Acquiring new customers typically costs 5-25 times more than expanding relationships with existing ones.

Pricing that scales with value is fundamental to expansion. Ensure your pricing model allows customers to start small and pay more as they derive increasing value, whether through usage-based pricing, tiered features, or seat-based models (enterprise teams).

Creating natural upsell paths keeps expansion feeling organic. The best expansions feel like natural progressions rather than sales tactics. As users become more sophisticated, they should naturally encounter needs that your premium offerings address.

Cross-selling complementary products leverages established trust. Once you've established trust in one area, you can leverage that relationship to solve other (and possibly more important) problems for your customers.

When I present expansion plans to executives who are fixated on new logos, I remind them that revenue is revenue. A dollar from an existing customer is actually worth more than a dollar from a new one, because it comes with lower acquisition costs and typically higher margins.

Example: Benny Brew Beans

Benny Brew Beans started as a modest coffee subscription service that delivered specialty beans to customers' doors. Their initial growth was slow but steady, primarily through word-of-mouth referrals from coffee enthusiasts.

When Paul took over as their product head, he realized they needed a more structured approach to growth. He began by focusing on product-market fit, confirming they had something customers truly wanted through customer interviews and usage data. Their retention curves had flattened around 70%, and they noticed subscribers became genuinely upset when deliveries were delayed — both great signs!

Once confident in their product-market fit, Paul implemented a framework based on the four pillars of sustainable growth:

For acquisition, they shifted from expensive Instagram ads to a referral program where existing customers could gift a free sample box to friends — a much more cost-effective approach that brought in customers with similar profiles to their best current users. They tracked channel-specific CAC (Customer Acquisition Costs) and found referred customers cost 60% less to acquire than those from paid social.

For activation, they discovered their "Aha! Moment" came when customers brewed their first cup using the included brewing guide. They streamlined the onboarding process, introducing a simple app that walked users through optimal brewing methods for their specific beans. Time-to-value dropped from five days to just one.

For retention, they launched a "Coffee Explorer" program where subscribers received progressively more adventurous beans based on their taste preferences. Customers who participated in this program had a 95% renewal rate compared to 65% for those who didn't. They also added tasting notes and origin stories to each delivery, creating emotional connections to the products.

For expansion, they introduced a natural upsell path. Once customers had been subscribers for three months, they were offered premium equipment like grinders and brewers that complemented their coffee preferences. Their cross-sell attachment rate reached 35%, and average revenue per user grew by 40% within six months.

Within a year, Benny Brew Beans' revenue had doubled, but more importantly, their business model became sustainable with a healthy 3:1 LTV (Lifetime Value) ratio. Paul's team focused on the metrics that mattered — cohort retention, second-order revenue, and expansion percentage — rather than vanity metrics like total sign-ups.

The coffee was great, but their growth approach was pipin' hot! Go, Paul!

Measuring What Matters

Now that we've covered the four pillars of sustainable growth, let's talk about measurement. The metrics, on which you choose to focus, will shape your team's behavior and ultimately determine your growth trajectory.

I've seen too many products fail because they optimized for vanity metrics. Those are typically numbers that look good in presentations but don't translate to any kind of business value. Let's look at the metrics that actually matter for each pillar of growth.

Acquisition Metrics That Matter

Blended CAC and channel-specific CAC help you understand how much you're spending to acquire each customer, in total and by channel. CAC payback period shows how long it takes to recover your acquisition costs. The CAC:LTV ratio reveals for every dollar spent on acquisition, how many dollars you get back over the customer lifetime. And your organic acquisition percentage tells you what proportion of new customers come through unpaid channels.

When you acquire a new customer, make sure your development partner tags their record with a CAC value and an identifier of how they were obtained (the funnel). Then, you can later run a SQL query to determine, for example, how much they've spent on your site or in your store and what that original acquisition cost was. The difference will allow you to spot much more valuable acquisition channels.

Activation Metrics That Matter

Time to value measures how long it takes for a new user to experience your product's core value. Activation rate tracks how long it takes new users to first use your product's core feature(s). First-session depth reveals how deeply users engage during their initial interaction. Second-day return rate shows whether users come back after their initial experience.

Like with acquisition metrics, make sure your tech partner is tracking core feature usage events. Without this type of event tracking, finding this information would be impossible.

Retention Metrics That Matter

Cohort retention curves help you see how retention behavior changes over time for different user groups. Net revenue retention reveals whether existing customers are generating more or less revenue over time. Feature adoption breadth shows how many of your key features the average user engages with. Engagement frequency tells you how often users return to your product, and whether this frequency is healthy for your use case.

As with the other metrics, being able to query your products' feature usage by existing, established customers is critical. If you don't know which features they rely on, build onto and augment those features, it will be hard to keep them from going to your competitor.

Expansion Metrics That Matter

Expansion revenue percentage reveals what proportion of new revenue comes from existing customers. Average revenue per user (ARPU) growth shows whether customers are paying you more over time. Upgrade rates measure what percentage of eligible customers choose to upgrade when offered. Cross-sell attachment rate demonstrates how successfully you're selling additional products to your base.

Beyond these pillar-specific metrics, three North Star indicators will give you a holistic view: Net dollar retention (NDR), Lifetime value (LTV), and Profitability per customer.

I've worked with product teams who celebrated hitting 10,000 signups in a month, only to discover that 9,800 of those users disappeared within weeks. We'd have been better served acquiring 1,000 users who stuck around. This is why I'm ruthless about focusing on metrics that connect to sustainable business value.

I've found that explaining the difference between vanity metrics and actionable metrics to some stakeholders is like trying to convince someone that their adorable but destructive puppy isn't actually a good guard dog. "But look how impressive those teeth are!" they say, while it's busy chewing through the foundation of the house!

Building a Growth-Oriented Organization

Sustainable product growth isn't just about strategies and metrics; it's about building an organization capable of executing consistently over time. Here's what I've found works...

Cross-Functional Growth Teams

The most effective growth initiatives typically require collaboration across product, engineering, marketing, sales, and customer success. Rather than keeping these functions siloed, consider creating dedicated cross-functional growth teams organized around specific user journeys or business outcomes.

Individuals or teams should be put to task on new user onboarding and activation, monetization and revenue expansion, core engagement and retention, or virality and referral programs. Each team should have clear metrics they're responsible for moving, along with the autonomy to run experiments and implement changes within their domain.

Experimentation Culture

Sustainable growth requires continuous learning and adaptation. Build systems that enable rapid, rigorous experimentation through hypothesis-driven development, experiment tracking, minimum viable tests, and measuring learning velocity.

I once consulted with a team that proudly ran 50+ experiments per quarter, yet couldn't tell me what they'd learned or how those results influenced their roadmap. Meanwhile, a competitor ran just 10 experiments but used each one to fundamentally improve their understanding of the customer, ultimately overtaking my client in market share. My advice: create an internal Confluence or Notion site which publishes your frequently updated metrics and research findings for all across your organization to see.

Data Infrastructure

Growth decisions should be data-informed (though not exclusively data-driven). Invest in data infrastructure that provides user behavior tracking, cohort analysis capabilities, experimental methods, and predictive analytics.

I worked with one organization that invested heavily in a massive event logger and database, before they had clear questions they needed to answer. They ended up with a costly system that provided endless reports but few actionable insights.

Start with the decisions you need to make, then work backward to determine what data and systems are required. Work closely with your technical partner(s) and determine which systems can efficiently get the job done, allow you to have easy insight into your data, and whereby the customer isn't impacted by the event logging.

Example: Rockwhite Analytics

Rockwhite Analytics was a company that transformed their approach to product growth by embracing both experimentation culture and robust data infrastructure.

When Dari took over as Director of Product, Rockwhite was struggling with flat user numbers despite having decent product-market fit. Their product dashboard helped marketing teams track campaign performance, but growth had stalled.

Dari noticed that product decisions were based mostly on gut feeling and HiPPO (Highest Paid Person's Opinion). Team meetings involved passionate debates about features, but rarely included concrete user data.

Her first move was establishing a simple experiment. She introduced "Test Tuesdays" where the team would formulate clear hypotheses about user behavior. For instance, they hypothesized that simplifying the reporting workflow would increase daily active usage.

Each experiment had a clear structure: the hypothesis, success metrics, minimum sample size, and duration. The team documented everything in shared Notion pages, making results transparent across departments.

Before running experiments, Dari worked with engineering to revamp their data infrastructure. They implemented proper event tracking to capture user journeys through the product. Rather than logging everything possible, they focused on key actions that aligned with business outcomes. They tracked how often reports were generated, dashboard sharing frequency, and feature adoption rates.

This focused approach to data collection meant engineers weren't overwhelmed building complex systems, and analysts could quickly extract meaningful insights.

The combination worked wonders. When they tested a redesigned onboarding flow, the data showed a 32% improvement in new user activation. When they experimented with in-app guiding tooltips, usage of previously ignored features increased by 18%.

The team started making decisions based on evidence rather than opinions. Dari insisted they build a learning library, where experiment results were shared throughout the company. Failed experiments were celebrated as valuable learning opportunities rather than mistakes.

Within a year, Rockwhite had run over 40 experiments. The insights gained helped them prioritize their roadmap based on proven user behavior rather than assumptions. Monthly active users grew 40%, and customer retention improved dramatically — all because they built a culture of experimentation backed by just enough data infrastructure to make informed decisions.

Customer-Centric Culture

Finally, sustainable growth comes from deeply understanding your customers and building for their evolving needs. Foster a customer-centric culture through regular customer interaction, shared customer insights, customer outcome metrics, and customer journey mapping.

At one large software company, we implemented a usability lab, where we recorded and shared short video clips of customers describing their challenges and how our product helped (or failed to help) them. These unfiltered perspectives became a powerful tool for building empathy and guiding product decisions.

I've found that the teams most resistant to customer feedback are often the ones most in need of it. It's like someone walking around all day with spinach in their teeth, insisting they don't need a mirror. The longer they avoid looking, the more embarrassed they are hours later.

Growth Pitfalls to Avoid

In my years observing product growth, I've observed several common pitfalls that can derail even well-intentioned growth efforts.

Premature Scaling

As mentioned earlier, pushing growth before establishing product-market fit is a recipe for wasted resources. Signs you might be scaling prematurely include high acquisition costs that aren't decreasing over time, poor retention metrics beyond the first month, low Net Promoter Scores (NPS) or customer satisfaction, and difficulty articulating your ideal customer profile.

I've seen startups burn through millions in funding to acquire users for products that weren't ready, only to watch those users never return. Be honest about where you are in the product-market fit journey, and focus your resources accordingly.

Channel Dependency

Becoming overly reliant on a single acquisition channel puts your growth at risk. Whether it's SEO, paid social, partnerships, or app store features, external channels can change their algorithms, pricing, or policies without warning.

A friend's very successful consumer SaaS company had built their entire acquisition strategy around Facebook ads. When iOS 14.5 privacy changes disrupted their targeting capabilities, their CAC increased by 40% overnight. They had no alternative channels developed and faced an existential crisis.

Diversify your acquisition channels early, even if it means accepting some inefficiency in the short term. The resilience this provides is worth the investment.

Growth Without Unit Economics

Growth that doesn't translate to profitability is ultimately unsustainable. I've seen too many teams celebrate user or revenue growth without understanding whether they're actually creating or destroying value with each new customer.

What is your fully-loaded cost to acquire a customer? What is your gross margin on the product or service? How long does a customer need to stay to become profitable? Are retention rates high enough to reach profitability? These questions need answers.

One company I worked with was celebrated for growing ARR by 300% year-over-year. However, a deeper analysis revealed they were losing money on each customer for the first 14 months, and their average customer lifetime was only 11 months. They were effectively paying customers to use their product, creating the illusion of growth while destroying value.

Feature Bloat

In the pursuit of growth, there's a temptation to continuously add features to address specific customer requests or market segments. This often leads to product complexity that undermines the core experience.

I worked with one product team that added tons of new features in only a few sprints, proudly reporting that they had addressed the top customer requests. Six months later, their NPS had dropped significantly. User interviews revealed that while people had asked for these features, the resulting complexity made the product less usable overall.

Sometimes the best growth strategy is to do fewer things better. Focus on the core value proposition that made customers choose you in the first place, and be extremely selective about what you add beyond that. Additionally, always test a small bundle of new features to a small customer segment, before bestowing all of it on your entire user base.

Ignoring the Customer Experience Holistically

Growth isn't just about product features; it's about the entire customer experience from first touch to long-term engagement. I've seen teams obsess over in-product metrics while ignoring critical touch points in the broader journey.

For example, one B2B product I advised had excellent in-product engagement metrics but struggled with retention. A little bit of research revealed that their onboarding process was smooth, but customers weren't receiving adequate support during implementation. The product itself worked well, but customers weren't successfully integrating it into their workflows.

Map your customer journey end-to-end, and measure satisfaction at each stage. Often, the biggest growth opportunities lie in the transitions between stages rather than within the product itself.

My favorite analogy for holistic customer experience is the restaurant industry. You can have the best chef in the world, but if the reservation system is frustrating, the seating is uncomfortable, or the bill is incorrect, customers won't return. Every touch point matters.

When you rollout a new feature, get a customer on a video chat and screen share and walk with them through using it. Don't navigate them. Watch if the usage is intuitive and fluid or if they stumble. Don't make assumptions without seeing the journey in real-time.

The Future of Product Growth

As we look ahead, several trends are reshaping how we should think about sustainable product growth.

Privacy-First Growth Strategies

With increasing privacy regulations and platform changes, the era of unlimited user tracking and hyper-targeted acquisition is ending. Forward-thinking growth teams are building first-party data strategies that provide value to users, developing contextual targeting approaches that don't rely on personal data, creating compelling reasons for users to opt into data sharing, and focusing on channels they directly control (like email and in-product messaging).

Community-Led Growth

Communities are emerging as powerful growth engines, particularly for products with learning curves or collaboration components. Community members not only provide peer support but also create complementary content and extensions, evangelize the product to new potential users, provide invaluable product feedback and use cases, and reduce churn through social connections.

I've seen products with strong communities achieve CAC rates 60-70% lower than industry averages, while maintaining higher retention rates.

There are many reasonably priced, white-labeled, easily branded, community SaaS platforms you and/or your technical partner(s) can put into service. One which immediately comes to mind is Hivebrite.

Ecosystem Thinking

The most sustainable growth strategies now extend beyond the product itself to encompass broader ecosystems. This includes developer platforms and APIs that enable integrations, marketplace dynamics that create network effects, partner ecosystems that expand distribution channels, and complementary products that solve adjacent problems.

Companies that position themselves as platforms rather than just products often achieve more defensible market positions and more efficient growth.

Conclusion

Sustainable product growth isn't a campaign or a hack; it's a methodology built on a thorough understanding of customers, market dynamics, and unit economics. It requires patience, discipline, and a desire to make choices that may constrict short-term growth in order to create long-term value.

I hope the methods and lessons I've shared in this chapter help you build products that reach rapid growth and lasting impact. The market doesn't need more shooting stars that burn brightly and burn out quickly. It needs those that shine consistently and act as a guide for years to come.

Let's move onto a short example, which drives home the concepts of this chapter...

Concepts in Action: FlavorVault

FlavorVault began as a simple recipe organization app for home cooks. Founded by culinary school dropout Toni Sharp and software engineer Manny Hendricks, the app allowed users to store recipes, generate shopping lists, and track ingredients in their pantry. Like many startups, they faced a crucial question: how could they turn their small but passionate user base into a thriving business?

Product-Market Fit

FlavorVault's early versions had decent download numbers but lackluster retention. Users would try the app once or twice before abandoning it. The team was tempted to ramp up marketing spend, but Toni remembered advice from her former mentor: "Scaling something that doesn't work just creates a bigger version of something that doesn't work."

Instead, they spent six months interviewing users and iterating on their product. The breakthrough came when they discovered their most loyal users weren't just storing recipes; they were adapting them based on dietary restrictions and ingredient availability. This insight led to FlavorVault's killer feature: an AI-powered recipe modifier that could transform any recipe to accommodate allergies, diets, or missing ingredients.

When they released this feature, something remarkable happened. Users began sharing modified recipes with friends and family, creating natural word-of-mouth growth. As Toni told her team, "We've found our 'Aha!' moment. People aren't just organizing recipes; they're solving daily cooking frustrations."

Thoughtful Acquisition

Rather than targeting all food enthusiasts, FlavorVault narrowed their focus to parents managing family meals with dietary restrictions; users with the highest lifetime value. While competitors poured money into broad Instagram campaigns, FlavorVault partnered with allergy-friendly food bloggers and specialized in SEO keywords like "gluten-free substitutions" and "kid-friendly vegan meals."

Their "Recipe Rescue" feature — allowing users to snap photos of cookbook recipes and instantly modify them — became particularly popular among this demographic, driving organic growth through specialized Facebook groups and forums.

"Acquisition isn't about getting the most users," Toni explained to her investors. "It's about finding the right users. We're not throwing a party and paying people to attend; we're creating an event so valuable that people bring their friends."

Meaningful Activation

FlavorVault's data revealed a critical insight: users who successfully modified a recipe within their first three days were 4-times more likely to become paying subscribers.

The team redesigned their onboarding flow to guide new users directly to this value. Rather than overwhelming newcomers with all of FlavorVault's features, they focused on one clear goal: help users successfully modify a recipe within minutes of downloading the app.

They implemented a "Quick Start" option where users could modify a simple pasta recipe to suit common dietary needs. This approach reduced their time-to-value from 2.5 days to under 10 minutes.

"Good activation is like teaching someone to ride a bike," Manny told the development team. "We need to hold the seat just long enough for them to feel the thrill of balancing on their own."

Deliberate Retention

Analysis showed that FlavorVault users who engaged with at least four of their core features had renewal rates above 85%, compared to just 35% for those using fewer features.

The team launched a "Feature of the Week" campaign, highlighting underutilized tools through in-app notifications and email tips. They created integration points between features. The meal planner would suggest using the pantry tracker, which would then recommend the shopping list generator.

Most importantly, they studied which features created habitual usage. Their "Weeknight Wizard" tool, which suggested quick meals based on available ingredients, became a Monday night ritual for many users. They enhanced this feature with timely reminders and personalized suggestions based on past behavior.

Strategic Expansion

With a solid retention strategy in place, FlavorVault explored thoughtful expansion opportunities. Rather than simply raising prices, they introduced a tiered model that aligned with increasing value:

- **Free**: Basic recipe storage and limited modifications
- **Gourmet** ($5.99/month): Unlimited recipe modifications and advanced meal planning
- **Chef's Table** ($12.99/month): Added AI-generated recipes, cooking technique videos, and integration with smart kitchen appliances

Each tier represented a natural progression in the user journey. As home cooks became more confident with basic modifications, they naturally wanted the additional features offered in higher tiers.

The team also introduced a premium offering for small restaurants and catering businesses, allowing them to scale recipes and calculate food costs; an adjacent market opportunity that leveraged their existing technology.

Measuring What Matters

FlavorVault initially celebrated vanity metrics like downloads and registered users. However, Toni instituted a rigorous focus on actionable metrics for each growth pillar:

Acquisition Metrics:

- Channel-specific CAC with special attention to organic acquisition percentage
- CAC ratio by user segment (which revealed that parents with multiple dietary restrictions had 3x the lifetime value of general food enthusiasts)

Activation Metrics:

- Time to first recipe modification (target: under 10 minutes)
- Percentage of new users who modified at least one recipe in first week

Retention Metrics:

- Monthly active users who engaged with 4+ features
- Weekly usage patterns (identifying the Monday "Weeknight Wizard" ritual as their retention anchor)

Expansion Metrics:

- Upgrade rates from free to paid tiers
- Average revenue per user over time

The team created a "Metrics Wall" in their office and online dashboard, ensuring everyone knew which numbers mattered and how they were trending.

Building a Growth-Oriented Organization

FlavorVault reorganized their team structure around key user journeys rather than functional departments. Cross-functional pods focused on new user activation, core engagement, and monetization. Each pod included product, engineering, design, and marketing team members, with clear ownership of specific metrics.

They established a rigorous experimentation culture, running weekly A/B tests on critical user paths. Unlike competitors who ran dozens of unfocused experiments, FlavorVault concentrated on high-impact tests that answered specific questions about user behavior.

"We're not running experiments to look busy," Toni told her team. "We're running them to learn something important about our users."

Growth Pitfalls Avoided

FlavorVault sidestepped several common growth traps:

- **Premature Scaling**: Instead of pushing growth before achieving product-market fit, they invested in user research and product refinement, ensuring they had something worth scaling.

- **Channel Dependency**: While Facebook groups initially drove significant growth, the team developed multiple acquisition channels, including SEO, partnerships with nutritionists, and a referral program. This diversity proved valuable when Facebook changed its algorithm, temporarily reducing visibility in groups.

- **Feature Bloat**: Despite user requests for advanced recipe creation tools, the team remained focused on their core value proposition of recipe modification and meal planning. They introduced a voting system for new features, ensuring they only built what truly enhanced the core experience.

- **Ignoring Holistic Experience**: When data showed users struggling to import recipes from certain websites, the team didn't just improve the importer; they created a human-powered recipe input service for premium users, addressing a crucial pain point outside the core product.

The Future: Community and Ecosystem

As FlavorVault matured, they recognized the power of community-led growth. They launched "FlavorVault Collective," a platform where users could share modified recipes and meal plans. This community not only provided value to users but also generated a wealth of content that improved their SEO and reduced support costs.

The team also embraced ecosystem thinking, developing an API that allowed kitchen appliance manufacturers to integrate with FlavorVault. Smart ovens could automatically adjust cooking temperatures based on recipe modifications, while grocery delivery services could fulfill shopping lists with a single click.

"We're no longer just an app," Toni told her team during their fourth anniversary celebration. "We're building a cooking ecosystem that solves real problems for people with dietary needs. And the best part? We're growing like a continent; slowly perhaps, but with unstoppable momentum."

Execution Essentials

Foundation: Product-Market Fit & User Acquisition

- Product-market fit exists on a spectrum. Aim for the point where your product becomes the obvious solution to a meaningful problem for a specific audience.

- Look for qualitative signals: customers becoming upset at the thought of losing your product is a powerful indicator of true product-market fit.

- Target the right users who match your ideal customer profile rather than pursuing all potential users.

- Build sustainable acquisition loops that become more efficient over time, not less.

- Tag customer records with acquisition channel and CAC values to identify your most valuable acquisition sources.

Value Delivery: Activation & Retention

- Identify your product's core "Aha! Moment"; the key action or experience that correlates with long-term retention.

- Streamline the path to value by removing unnecessary friction between signup and first value experience.

- Implement "usage depth" initiatives; users who engage with more features typically have significantly higher retention rates.

- Study and enhance features that create habitual usage patterns among your most loyal customers.

Revenue Growth: Expansion & Measurement

- Design pricing that scales with value, allowing customers to start small and pay more as they derive increasing benefits.

- Create natural upsell paths that feel like organic progressions rather than sales tactics.

- For acquisition, track blended CAC, channel-specific CAC, CAC payback period, CAC ratio, and organic acquisition percentage.

- Focus on three North Star indicators for holistic view: Net dollar retention (NDR), Lifetime value (LTV), and Profitability per customer.

- Remember that revenue from existing customers is worth more than revenue from new ones due to lower acquisition costs.

Organizational Excellence & Future Growth

- Create cross-functional growth teams organized around specific user journeys or business outcomes.

- Watch for signs of premature scaling: high acquisition costs that aren't decreasing, poor retention beyond the first month, and low customer satisfaction.

- Develop privacy-first growth strategies that provide value while respecting increasing privacy regulations.

- Invest in community building as communities can drive significantly lower CAC and higher retention rates.

- Adopt ecosystem thinking by creating platforms, APIs, and marketplaces that extend beyond your core product.

The Product
Chapter 13

Smart Monetization

I want to dive deep into an aspect of product growth that deserves special attention: monetization. After all, sustainable growth ultimately requires sustainable revenue.

In my experience, monetization is where art meets science in product management. It requires both analytical rigor and intuitive understanding of human psychology. Get it right, and you create a virtuous cycle where revenue fuels improvements that increase willingness to pay. Get it wrong, and you can sabotage even the most remarkable product.

The Psychology of Value Perception

At its core, monetization is about value exchange. Your customers exchange money for the value your product provides. But here's the critical insight: value isn't objective; it's perceived. And perception can be influenced.

The perceived value of your product is shaped by numerous factors beyond the product itself. Market positioning, competitive alternatives, brand equity, purchase context, and even pricing presentation all play crucial roles in how much value customers believe they're receiving.

I once worked with a massively popular personal database application that was struggling with monetization despite strong usage metrics. Our research revealed that users loved the product but didn't think their productivity gains were due to the tool itself.

They attributed their success to their own improved habits. The value was there, but perception wasn't. So, the company bundled it with an office productivity suite.

The team redesigned the experience to periodically quantify and celebrate the time saved and goals achieved, directly attributing these wins to the app's features. Within three months, willingness to pay for the app as a standalone tool increased by over 50% without a single change to the core product functionality.

This experience taught me that effective monetization isn't just about extracting value but making value visible and tangible. Some of the most effective tactics for increasing perceived value include:

Value Anchoring

Contextualizing your offering against more expensive alternatives sets a psychological reference point. If you manage a media analytics product, increase your conversion rate by simply showing the cost of hiring an analytics consultant alongside your subscription price.

Results Visualization

Help your users see the concrete outcomes of using your product. If you manage a project management tool, implement a "Time Saved Dashboard" that quantifies hours saved through automation and streamlined workflows, which dramatically improves renewal rates.

Concrete Success Metrics

Tie your product directly to business outcomes that matter. If you manage an email marketing platform, create a "Revenue Generated" calculator that estimates the direct financial impact of campaigns sent through your system, allowing you to charge much higher rates.

The key is to not just create value, but to ensure that users consciously register and attribute that value to your product. This forms the foundation of any successful monetization strategy. It's rather like being a magician who actually has to point out their own tricks. "Did you see that rabbit I just pulled out of my hat? It's worth at least $29.99 a month, wouldn't you agree?"

I often tell my teams that monetization is like oxygen; it needs to be present throughout the entire product experience, not just when the user is engaging in a transaction. Every use of a feature is an opportunity to reinforce your product's value to the user.

Example: ProductivityPro

ProductivityPro was a task management app that had amazing user engagement metrics. People were using it daily to organize their work and personal projects, but when they tried to convert their free users to the premium tier, the numbers were disappointing.

The product team realized their core issue: users loved the app but didn't connect their productivity improvements to the tool itself. Users thought they were just "getting better at organizing" rather than benefiting from ProductivityPro's features.

So they made some key changes. They added a "time saved dashboard" that showed users exactly how many hours they'd saved by using specific ProductivityPro features compared to their previous methods. For instance, "Your quick-add shortcut saved you 45 minutes this week" or "Template sharing has reduced your meeting prep time by 70%."

They also implemented periodic celebration moments — when users completed projects ahead of schedule, a notification would highlight how ProductivityPro's kanban visualization had helped them spot bottlenecks early.

Within three months of these changes, their premium conversion rate jumped from 3% to nearly 15%. The product hadn't fundamentally changed — they had simply made the value visible and helped users attribute their success to the app rather than just to themselves.

What worked was connecting concrete outcomes to specific features. Users started to recognize that their productivity gains weren't just personal improvement but were directly enabled by the tool they were using. When renewal time came around, the decision to pay became much easier because users could see the tangible benefits they'd receive.

Designing Your Monetization Model

The structure of your monetization model can be as important as the price itself. The right model aligns incentives between you and your customers, scales with value delivered, and creates predictable revenue for your business.

Here are the primary monetization models I've implemented across different products, along with when each tends to work best:

Subscription Models

Having a subscription model creates predictable, recurring revenue and tends to work well for products that provide ongoing value. The key to subscription success is demonstrating continuous value delivery to prevent churn.

For an online community SaaS product I led, we initially struggled with monthly renewals despite strong initial adoption. Investigation revealed that usage typically spiked during website redesigns but declined during maintenance periods. Users questioned the value during these quiet periods.

Our solution was to introduce fresh website designs, new ways of smiling at other members, and simple ways to filter only those members of interest. These touchpoints reminded customers of the continuous value they received even during lower-usage periods, thus increasing renewal rates.

Usage-Based Pricing

This model aligns perfectly with value when consumption directly correlates with benefit received. It allows users to start small and grow their spending as they derive more value.

For products such as APIs, shift from tier-based pricing to a pure usage model with volume discounts. This will eliminate the "cliff effects" between tiers that cause customer frustration. This will not only improve customer satisfaction but increase average revenue per account, as customers will become comfortable using more of the service without fear of hitting the next pricing tier.

Freemium Models

Freemium products combine a free basic version with premium paid features. They work best when your product has strong network effects or when the free version serves as effective marketing for paid tiers.

One mobile app team I advised was struggling with their freemium conversion rates. Analysis revealed that their free/paid boundary was placed at a feature (advanced formatting) that users rarely discovered before deciding whether to pay. The team redesigned the model to place the boundary at team size instead (free for up to 3 users, paid beyond that), which aligned perfectly with their value proposition of team coordination. Conversion rates tripled within 60 days.

Marketplace Transaction Models

Taking a cut of transactions facilitated through your platform helps to align your success directly with your users' success. This tells them that it's in your best interest to ensure they also succeed.

If your product is a platform for freelancers and gig work, a flat 15% fee might be discouraging for high-value transactions. Instead, implement a sliding fee structure that decreases as transaction value increases, which incentivizes premium listings. The result is likely to be a substantial increase in gross billing value with only a slight decrease in the average commission rate. Everyone wins.

Enterprise Licensing

For larger organizations, negotiated contracts might be best. This model works for products that deliver organization-wide value or require significant customization.

For a popular corporate billing tool I once developed, I initially resisted enterprise licensing in favor of per-seat developer pricing. However, I found that seat-based models created perverse incentives for customers to limit user access, reducing overall product value and engagement. By switching to an enterprise model with unlimited users but tiered based on data volume, we both increased our average contract value and improved product adoption within customer organizations.

What happened in the year to follow was a nice surprise. I ended up having more developers using my product. When they changed jobs, they brought along that product knowledge and loyalty. More contracts (and revenue) were born as a result.

The most sophisticated products often combine multiple models. For instance, a collaboration tool might offer a freemium model for small teams, usage-based components for storage or integrations, and enterprise licensing for organizations above a certain size.

The key is ensuring that your monetization model grows in alignment with the value customers receive. When customers feel they're getting more value as they pay more, price resistance decreases dramatically.

If you're running a lemonade stand, you can simply charge by the cup. But building sustainable product growth requires monetization models that scale with value, align incentives, and adapt to changing market conditions. Trust me, I've watched enough executives try to apply lemonade stand pricing models to SaaS businesses to know the difference. It's like watching someone try to pay their mortgage with a credit card. Ain't gonna happen.

Executing Your Monetization Strategy

Even the best-designed monetization strategy can fail in execution. Here are the implementation considerations I've found most critical to monetization success:

Timing and Sequencing

This can dramatically impact monetization outcomes. Early introduction of pricing can work for serious users but may limit viral growth. Late introduction risks training users to expect free value.

If your product is a design collaboration tool (such as Figma), delay monetization until you reach a certain amount of active users – let's say

100,000 – and focus entirely on product-market fit. When you then introduce pricing, grandfather existing users onto generous free plans while implementing freemium for new users. This approach will preserve your current network while beginning to generate revenue from those who derive the most value.

Another tactic is to introduce paid plans from day one but with an extended free trial. This approach works for serious business users and establishes value perception early, resulting in higher average revenue per user but slower initial growth. This tactic also avoids pissing off the loyal users who brought you to the dance. No surprises!

There's no universal right answer on timing. It all depends on your growth strategy, funding situation, and current market dynamics. The key is making a deliberate choice rather than relying on conventional wisdom.

Sales Channel Alignment

Ensure your monetization model works with how you sell. Different sales techniques require different pricing approaches:

- Self-service products need transparent, simple pricing that facilitates quick decisions.

- Sales-assisted approaches can support more complex or customized pricing.

- Channel or partner sales typically require margin structures and deal registration.

If you're not working with a very costly enterprise product, stick to a self-service method of pricing. When a user feels they're in control of the transaction, they also feel they're getting a better shake.

Testing and Optimization

Don't treat monetization like it's not an ongoing process or that it's a one-time decision. Some tasks you should be performing regularly are:

- A/B testing of pricing pages and checkout flows.

- Team meetings and analysis of monetization performance over time.

- Offers and promotions for various customer segments, such as regional, industry, or those with enabled capabilities.

- Controlled price increase tests with existing customers.

Attempt to run continuous price testing with new users, gradually increasing prices as you validate willingness to pay. You might discover that your team is leaving a lot of revenue on the table and your customer's LTV (Lifetime Value) is lower than it could be.

Communication Strategy

Have a good and thorough communication strategy will show how customers perceive your monetization approach. Without one, you're going to frustrate even the tamest customers. Effective communication includes:

- Clear articulation of your value proposition.
- Transparent explanation of pricing structure.
- Advance notice of any price changes.
- Consistent messaging across marketing, sales, and product.

For a small developer tools company I advised, we needed to implement a significant price increase to support continued product development. Rather than simply announcing higher prices, we created a comprehensive communication plan that explained the specific investments we would make with the additional revenue, outlined the new capabilities customers would receive, and provided a generous grandfathering period for existing users. The result was 93% retention through the price change, far exceeding industry benchmarks for similar increases.

The implementation of your monetization strategy should be treated with the same care as the strategy itself. A brilliant pricing model poorly executed will underperform a decent model executed with excellence. It's like having a Michelin-star recipe but cooking it in a microwave. You might have all the right ingredients, but your dinner guests aren't going to like the outcome!

I've often found that product leaders approach pricing changes with trepidation, fearing customer backlash. But in my experience, customers are far more reasonable than we give them credit for. If you've built a product that delivers genuine value, communicated that value effectively, and designed a fair exchange, most customers will accept reasonable monetization approaches. Those who don't might not be the customers you want to build your business around anyway.

Good pricing communication is the essence of why I coach aspiring PMs on always providing honest transparency.

Example: DesignDeck

DesignDeck, a collaborative UI/UX tool for remote design teams, launched in 2023 with a freemium model that wasn't performing well. Their conversion rate hovered around 2%, and customer feedback showed confusion about their value proposition.

Their product lead, Craig, realized they had a timing and communication problem. They'd introduced pricing too early without properly demonstrating value, and their pricing page was overly complex.

Craig implemented three key changes. First, they extended their free trial from 14 to 30 days, giving users time to experience real value. Second, they simplified pricing from five complicated tiers to three straightforward options based on team size. Finally, they created a comprehensive communication strategy that highlighted specific design problems their tool solved and included short testimonial videos from current customers.

When they needed to increase prices six months later, they gave existing customers three months' notice and explained exactly how the additional revenue would fund specific new features they'd requested. They also grandfathered loyal early adopters at their original price point for an additional year.

The results were impressive — conversion rates jumped to 7%, customer retention improved by 35%, and they maintained 91% of their customer base through the price increase. By treating monetization as a core product discipline rather than just a finance function, DesignDeck transformed their business trajectory.

Monetization Maturity

Just as your product evolves through its lifecycle, your monetization approach should mature over time as well. I've observed distinct stages of monetization maturity across many products:

Stage 1: Initial Monetization

At this stage, the focus is on validating willingness to pay and establishing base pricing. Simple models with minimal segmentation are typical. The key metrics are conversion rate to paid and average revenue per user.

With past products, I implemented a straightforward freemium model with a single paid tier. This simplicity allowed me to validate market demand without complicating the user experience or our internal operations.

Stage 2: Segmentation and Expansion

As your understanding of customer segments (geographical, industry, use case, etc.) deepens, you can introduce more targeted pricing and packaging. Multiple tiers, add-ons, or industry-specific bundles emerge. Key metrics expand to include tier distribution and upgrade rates.

If you managed a data integration platform, you might evolve from a one-size-fits-all model to three distinct tiers based on connection volume, plus add-on packs for specific industry connectors. This may allow you to capture more value from enterprise customers while remaining accessible to smaller businesses.

Stage 3: Value Alignment

At this stage, you develop sophisticated understanding of your value metrics and align pricing directly with value received. Usage-based components, success-based pricing, or ROI-driven models become feasible. Key metrics include expansion revenue percentage and net dollar retention.

If your product is a customer support and ticketing platform, try shifting from seat-based pricing to a hybrid model that incorporates both seats and resolution volume. This might better align with the value you product delivers (efficient resolution of customer issues) and create natural expansion as your customers' businesses grow.

Stage 4: Ecosystem Monetization

The most advanced stage involves monetizing the ecosystem around your core product. This might include marketplaces, developer platforms, data monetization, or certified partner programs. Key metrics include ecosystem revenue percentage and partner satisfaction.

Let's say your product is an enterprise workflow platform. You could create a developer marketplace where third-party developers could sell integrations and extensions. In turn, you might take a 20% commission on these sales, creating a new revenue stream that ultimately grows to represent a significant amount of total company revenue – while significantly enhancing the core product's value proposition.

Moving through these stages requires increasing levels of organizational sophistication, from basic billing capabilities to complex usage tracking, partner management systems, and marketplace infrastructure. I've witnessed organizations try to jump directly to stage 4 without mastering the basics. It's like watching someone attempt to compose a symphony when they haven't yet learned to play "Chopsticks" on the piano. Ambition is admirable, but some things simply have to be done in order.

I often compare monetization maturity to learning a language. In the beginning, you're focused on basic vocabulary and simple sentences (can we get someone to pay us at all?). As you progress, you develop more nuanced expression, idioms, and contextual understanding, until eventually you can engage in sophisticated dialogue that captures subtle distinctions in meaning and intent.

Monetization Pitfalls to Avoid

Through trial and error (sometimes painful error), I've identified several common monetization mistakes that undermine product growth:

The Complexity Trap

Overly complex pricing structures create cognitive load for customers and operational challenges internally. I've seen pricing models with so many variables that sales teams couldn't explain them and customers couldn't understand them.

Have you ever seen how you license and pay for any Microsoft enterprise, server, or database software products? It's so confusing and, in my opinion, so unnecessary. By the time it's over, you feel squeezed for every ounce of your already laughable budget.

The rule of thumb I've developed: if your pricing can't be explained clearly in under 60 seconds, it's probably too complex. Or as I once told a particularly enthusiastic pricing consultant who presented a 47-slide deck on our new "simplified" pricing structure: "If we need an FAQ to explain our FAQ, we've already lost."

The Discount Addiction

Excessive discounting not only reduces revenue but damages value perception. When discounts become expected, your actual price becomes meaningless.

One online learning platform I can think of has developed a culture where 90% discounts were standard and/or often hit or miss. Everyone is onto the game and they make every attempt circumvent paying full price for courses.

They would be far better off designing their pricing to better reflect actual selling prices, with much more modest discounts. This would reset customer expectations and increase perceived value as the product will no longer appear artificially expensive.

The Feature-Millstone Effect

Including too many features in base packages makes future monetization difficult. Features given away for free are nearly impossible to charge for later.

A project management product I once worked with had included advanced reporting in all plans from launch. When they later attempted to create a premium reporting tier, customer backlash was severe (including my own). They had to grandfather all existing customers permanently, significantly limiting the revenue potential of this new offering.

The Hidden Cost Oversight

Failing to account for the costs of serving different customer segments can lead to unprofitable growth. This is especially common with "land and expand" strategies.

Don't quickly celebrate landing several major enterprise customers through aggressive pricing. Do some cost analysis, which may reveal that those customers are consuming 8-10x more support resources than mid-market customers, making them unprofitable despite their larger contracts. Instead, implement a success fee structure for enterprise deployments that preserve margins without raising base prices.

The Deferred Monetization Delay

Waiting too long to implement monetization can create entrenched expectations of free value. This is the classic mistake of many consumer applications.

A content platform I advised had grown to millions of users before introducing any monetization. Their initial paid conversion rates were below 0.1%, as they had effectively trained users to expect the service for free. We ultimately had to completely rethink their model, focusing on creator tools rather than consumer subscriptions, as the original user base proved nearly impossible to convert.

Conclusion

The most fundamental mistake I've seen is treating monetization as a finance function rather than a core product discipline. Pricing is product. It defines the relationship between your solution and your customers, shapes user behavior, and ultimately determines the resources you'll have available for future development.

Years ago, I overheard someone comparing poor monetization to having a leaky fuel tank in a race car. You can have the best engineering, the most skilled driver, and perfect track conditions, but if you're leaking fuel, you'll still lose the race. Monetization is what fuels your product growth engine. It deserves just as much strategic attention as your product roadmap.

Concepts in Action: Zigler

Zigler is a cloud-based team collaboration platform that helps remote teams work together effectively. Founded by three former project managers who experienced firsthand the challenges of distributed work, Zigler combines video conferencing, document sharing, task management, and virtual whiteboarding in one seamless interface.

The Psychology of Value Perception

When Zigler launched, usage metrics were strong, but monetization was disappointing. Research revealed that while teams were spending an average of 12 hours per week in the platform, they didn't attribute their improved productivity to Zigler itself.

The product team implemented three key changes to make value more visible:

- Comparison chart showing the cost of using separate tools versus the all-in-one Zigler solution, highlighting annual savings of over $2,400 per employee
- Productivity dashboard that calculated hours saved through Zigler's integrations and automated workflows
- Collaboration ROI calculator that translated activity into business outcomes, such as "Your team completed 32 projects this quarter, a 28% increase from last quarter."

These changes increased willingness to pay by 65% within four months without altering core functionality.

Designing the Monetization Model

Zigler implemented a hybrid monetization approach to align with different customer segments. For small teams and startups, Zigler offers a free plan limited to 5 users with basic functionality, driving viral adoption while creating a natural upgrade trigger when teams grow. For established businesses, three subscription tiers provide increasing levels of functionality and support. To prevent perceived value drops between usage periods, they introduced monthly "Collaboration Insights" reports summarizing key metrics and achievements.

Advanced features like AI meeting transcription and analysis are charged based on usage, allowing teams to pay only for what they need while encouraging exploration of premium capabilities. For organizations with over 500 users, Zigler offers customized enterprise agreements with unlimited users but tiered based on total active minutes and storage needs, encouraging broader adoption.

Executing the Monetization Strategy

The company's execution strategy focused on timing, alignment, testing, and communication. Zigler launched with a completely free product for six months to build its user base. When they introduced pricing, they grandfathered early adopters to generous free plans while implementing freemium for new users, preserving their network while beginning to generate revenue.

For self-service customers (mostly small to mid-sized businesses), Zigler created transparent pricing with instant signup. For enterprise deals, they developed a consultative sales approach with demos and custom implementation planning. The team runs continuous A/B tests on pricing pages, checkout flows, and promotional offers. Their quarterly price sensitivity analysis has revealed opportunities to increase pricing for certain features without impacting conversion.

When the company needed to raise prices to fund new AI capabilities, they announced the change three months in advance. They detailed exactly what new features the price increase would fund and offered existing customers a one-year price lock. This approach resulted in 95% customer retention through the transition.

Monetization Maturity

Zigler's monetization has evolved through distinct stages. They started with a simple freemium model and a single paid tier at $10 per user per month, validating market demand without complicating operations. As they better understood customer segments, they introduced industry-specific packages. Their "Creator Studio" add-on for marketing teams and "Analytics Plus" for data-driven organizations allowed for more targeted value capture.

The company now incorporates meeting minutes, storage, and AI processing units in their pricing structure, directly tying costs to value received. This has increased net dollar retention to 118%. Most recently, they launched an app marketplace where third-party developers sell specialized tools and templates. With a 20% commission on these sales, the marketplace now represents 15% of total revenue while making the core product more valuable.

Avoiding Monetization Pitfalls

Zigler has carefully navigated common monetization traps. Their pricing is deliberately simple; any plan can be explained in three sentences, making purchasing decisions straightforward for customers. Rather than offering deep discounts, they provide value-adds like free onboarding or extra storage, preserving price integrity while still incentivizing purchases.

When introducing new capabilities, the team carefully considers which features belong in which tiers. Once they accidentally included advanced permission controls in all plans during a beta, making it impossible to later use as a premium differentiator. Zigler tracks support costs by customer segment. When they discovered that education customers required 3x more support but generated lower revenue, they created a specialized education team and adjusted pricing accordingly.

By treating monetization as a core product discipline rather than a finance function, Zigler has built a sustainable growth engine that aligns customer success with business success. Their thoughtful approach ensures they can continue investing in product improvements that deliver even more value over time.

Execution Essentials

Psychology of Value Perception

- Quantify time/money savings to make product value visible.
- Show comparisons to costlier alternatives for favorable reference points.
- Implement visualization features showing outcomes, not just features.
- Integrate monetization throughout the entire product experience.

Monetization Models

- Use subscriptions for ongoing value, but demonstrate continuous benefits.
- Implement usage-based pricing with volume discounts to avoid pricing cliffs.
- Design freemium boundaries aligned with your core value proposition.
- Use sliding fee structures in marketplaces to encourage premium transactions.

Execution Strategy

- Time monetization introduction based on target users and growth strategy.
- Match pricing complexity to sales channels (simple for self-service).
- Continuously test pricing pages, checkout flows, and promotions.
- Clearly communicate value gains when implementing price changes.
- Consider grandfathering existing users during significant pricing shifts.

Monetization Maturity

- Start simple to validate willingness to pay before adding complexity.
- Introduce segmentation based on customer needs as you grow.
- Align pricing with value metrics that scale with customer benefits.
- Explore ecosystem opportunities like marketplaces and partner programs.
- Master basics before attempting advanced monetization strategies.

Avoiding Pitfalls

- Keep pricing simple enough to explain in under 60 seconds.
- Limit discounting to maintain value perception and price integrity.
- Be strategic about which features to include in base packages.
- Implement monetization early to avoid expectations of perpetual free value.

The Product

Chapter 14

Mastering Data for Decision Making

The real skill in being a product leader lies in knowing what data to collect, how to analyze it meaningfully, and when to trust your experience over a spreadsheet.

In this chapter, I'll share what I've learned about turning the overwhelming ocean of available data into actionable product decisions. We'll walk through methods that have saved my teams from costly mistakes, metrics that actually predict success (rather than just looking impressive in presentations), and practical was of building a decision architecture that works in the messy reality of product development.

Whether you're drowning in data or struggling with too little, my goal is to help you cut through the noise and focus on what truly matters for your product's success.

Data Collection & Quality

The foundation of all data-driven decision making starts with the quality of your data. Even the most sophisticated analysis setup will lead you astray if it's built on faulty information.

Data Integrity

I'm sure you've heard of, "Garbage in, garbage out!". That old programmer saying has hung on for decades because it's always true, especially when it comes to product decisions. Data integrity refers to the accuracy, completeness, and consistency of your data throughout its lifecycle.

Early in my career, I worked on a product team that made a significant roadmap pivot based on what our metrics showed was a massive drop in usage of a particular feature. We spent three months rebuilding it, only to discover that the feature hadn't actually declined in usage at all. A gosh-darned tracking script had broken during a typical update. It taught me a valuable lesson about verifying data integrity before making major decisions.

Data integrity snafus typically arise from three sources: collection problems, processing errors, or interpretation mistakes. Collection problems include broken tracking links or setup, sampled data presented as more than it is, or biased collection methods. Processing errors happen when transformations corrupt data or when inconsistent time zones, currency conversions, or definitions create false patterns. Interpretation mistakes occur when we misunderstand what the data actually represents, like when the name of the database field is so abstract that it doesn't make sense (common, actually).

The best approach I've found is establishing regular data audits. At a minimum, have someone who didn't implement the tracking independently verify the data being collected. Create monitoring systems, of some sort, that alert you to sudden changes in key metrics, which often signal tracking problems rather than actual user behavior shifts. And document your data definitions meticulously. I've seen a few sprints of work wasted because different teams were using the same term to mean different things.

Collection Methods

Selecting the right data collection approaches determines whether you'll capture the insights that actually matter for your product decisions.

The collection methods you choose should directly connect to the questions you need to answer. For quantitative data, you'll typically rely on analytics platforms, A/B testing tools, and internal databases. For qualitative insights, user interviews, surveys, and usability studies become your primary sources. The key is matching your method to your specific inquiry rather than letting available tools dictate your approach.

One approach that has served me well is creating a tiered data collection strategy:

Tier One

Includes the minimum viable metrics you need to understand basic product health. Think of these as core conversion rates, retention, and essential engagement metrics.

Tier Two

Expands to include more granular behavioral data that helps diagnose problems identified in tier one.

Tier Three

Goes deep on specific features or workflows where you need detailed understanding. This tiered approach prevents the common problem of trying to track everything, which often results in tracking nothing effectively.

Remember that data collection should evolve with your product. Twitter (now X) originally tracked primarily tweet creation as their core metric, until they realized many users derived value simply from reading tweets without posting. Their collection methods evolved to capture this "invisible" engagement, fundamentally changing their understanding of their product. Similarly, your collection strategy should adapt as you learn more about how users actually get value from your product, not just how you initially thought they would.

Sample Sizing

Determining statistically significant data volumes is critical for confident decision-making, but it's an area where I see product teams frequently make expensive mistakes.

Sample size matters because it directly impacts the reliability of your conclusions. Too small a sample, and you're essentially making decisions based on anecdotes with numbers attached. Too large, and you're wasting resources chasing precision that won't meaningfully improve your decision. The appropriate sample size depends on several factors: the expected effect size (how big a change you're looking for), the baseline conversion rate, the confidence level you require, and the stakes of the decision.

I've found that many product teams default to arbitrary thresholds like "we need 1,000 users" or "let's run the test for two weeks" without considering whether these provide sufficient statistical power. Statistical power calculations aren't the most exciting part of product management, but they're worth understanding at a basic level. Tools like Optimizely's sample size calculator can help determine how much data you actually need before drawing conclusions.

A mistake I frequently observe is stopping tests as soon as they show statistical significance. This approach, called "peeking," dramatically increases your chances of false positives. Instead, determine your sample size in advance and commit to running the full test. Airbnb learned this lesson when they found that 80% of their product experiments that showed initial significance failed to maintain it when run to completion. They now use more rigorous statistical methods and predetermined sample sizes to avoid expensive product mistakes.

Qualitative vs. Quantitative Balance

Finding the right balance between hard metrics and user stories is perhaps the most sophisticated skill in a product leader's data arsenal.

Quantitative data tells you what is happening with statistical certainty, while qualitative data helps you understand why it's happening and how users feel about it. The most effective product decisions typically leverage both. As a rule of thumb, use quantitative data to identify patterns and opportunities, then qualitative data to develop hypotheses about the underlying causes and potential solutions.

Many teams fall into the trap of overvaluing one type of data at the expense of the other. Engineering-led organizations often dismiss qualitative feedback as "just anecdotes" while over-indexing on metrics. Design-led organizations sometimes make the opposite mistake, building solutions for problems articulated by a handful of vocal users without verifying the prevalence of those problems quantitatively.

Slack's early product development provides an excellent case study in balancing these approaches. Their quantitative data showed high engagement but didn't fully explain what made their product so sticky. Through qualitative interviews, they discovered users valued Slack as a unified "central nervous system" for their work, not just a chat tool. This insight informed their product strategy, positioning, and feature prioritization in ways that quantitative data alone couldn't have revealed.

I've found the most effective approach is maintaining separate "ledgers" for qualitative and quantitative insights, then regularly cross-referencing them to identify where they align and where they diverge. Points of alignment often represent your biggest opportunities, while divergence typically signals areas needing deeper investigation.

Analysis Methods

Having quality data is just the starting point. The analysis methods you apply determine whether that data transforms into actionable insights or remains just an interesting collection of numbers.

Cohort Analysis

Tracking distinct user groups over time reveals behavior patterns and retention trends that aggregate metrics often mask.

Cohort analysis separates users into groups based on their shared characteristics — most commonly their start date with your product — and tracks their behavior over time. This approach removes the distortion that comes from averaging across users at different lifecycle stages and exposes critical patterns in retention and engagement.

The power of cohort analysis lies in its ability to answer whether your product is actually improving over time. When Dropbox examined their retention by cohort, they discovered that despite growing overall user numbers, their core retention metrics weren't improving for new cohorts. This revelation led them to shift focus from acquisition to product improvements that would better retain users — a decision that ultimately strengthened their business fundamentally.

The most common cohort analysis mistake I see is looking only at time-based cohorts (users who joined in January versus February). While these are valuable, behavioral cohorts can yield even richer insights. Grouping users by their first action, feature usage, referral source, or demographic characteristics often reveals surprising patterns that can inform product strategy. For instance, analyzing users who first engage with different features can help identify which entry points correlate with long-term retention.

When implementing cohort analysis, start simple with just 2-3 cohort types and 3-4 key metrics before expanding. I've seen teams try to analyze dozens of different cohort combinations immediately, creating an overwhelming data swamp that yields no clear direction. Begin with time-based cohorts focusing on retention, then add one behavioral dimension that you believe might significantly impact user success.

Funnel Analytics

Mapping customer journey stages offers crystal-clear visibility into where your product experience succeeds or fails.

Funnel analytics break down multi-step processes into discrete stages, allowing you to identify exactly where users abandon your critical flows. The approach is straightforward: define the steps in your funnel, measure the conversion rate between each step, and focus optimization efforts on the points with the highest drop-offs.

The classic mistake in funnel analysis is defining your funnels too broadly or too narrowly. Too broad (like "visit → signup → purchase"), and you miss the nuanced friction points. Too narrow (tracking every micro-interaction), and you can't see the forest for the trees. I recommend starting with 4-7 meaningful steps for each critical user journey.

Discord discovered through funnel analysis that new users who joined servers with active moderators were significantly more likely to remain active. This insight wasn't visible in their aggregate metrics but became clear when they examined the conversion funnel divided by server characteristics. They subsequently prioritized features that helped server creators foster healthy communities, addressing a retention challenge they previously couldn't diagnose.

The most sophisticated funnel analysis goes beyond simply identifying drop-off points to understand why users abandon and what characterizes those who successfully convert. Segment your funnels by user attributes, traffic sources, or behaviors to identify which factors correlate with higher conversion. This approach transforms funnel analysis from a diagnostic tool to a predictive one.

A/B Testing

Systematic comparison of variations remains the gold standard for determining optimal product features and experiences.

A/B testing allows you to move beyond educated guesses by directly measuring the impact of changes on user behavior. The approach is deceptively simple: randomly assign users to different versions of your product and measure which performs better against your key metrics. The power comes from the scientific rigor this brings to product decisions.

The most common A/B testing mistake I encounter is running tests without clear, predetermined success metrics. "We'll see what improves" is not a testing strategy. Before launching any test, define your primary metric, secondary metrics you'll monitor for negative impacts, and the minimum meaningful improvement that would justify implementing the change.

Netflix's experimentation culture demonstrates the value of rigorous A/B testing at scale. Rather than relying on subjective opinions about what makes a good user experience, they test almost every change, from algorithm tweaks to UI elements. This approach led them to many counter-intuitive findings, like the discovery that showing fewer, more relevant recommendations outperformed showing more options — contradicting their initial hypothesis but ultimately improving user engagement.

One aspect of A/B testing that deserves more attention is the concept of interaction effects. Often, the impact of a change depends on other factors in the user experience. A feature that performs well for new users might perform poorly for power users. The solution is carefully designed experiment groups and segmented analysis. Start with broad tests to identify promising directions, then run more specific tests to optimize for different user segments.

Predictive Modeling

Using historical data patterns to forecast outcomes gives product teams a powerful advantage in anticipating user needs.

Predictive modeling applies statistical techniques to historical data to make informed forecasts about future behavior. While it sounds technically intimidating, even relatively simple models can provide valuable product insights without requiring a data science degree.

The key to effective predictive modeling is starting with clear business questions. Rather than asking "What can we predict?", ask "What would we do differently if we could predict X?" This ensures your modeling efforts connect directly to actionable decisions. Common applications include churn prediction, propensity to upgrade, and feature adoption likelihood.

Spotify provides an excellent example of predictive modeling applied to product development. Their Discover Weekly feature uses collaborative filtering and pattern analysis to predict which new songs a user might enjoy based on their listening history and similar users' preferences. The result feels almost magical to users but is based on systematic analysis of behavioral patterns.

For product teams new to predictive modeling, start with simple approaches before building complexity. Even basic cohort completion patterns can yield valuable predictions. If 80% of users who take actions A, B, and C within their first week become long-term customers, you can use this pattern to identify promising new users and tailor their experience accordingly. These simple models often provide the most actionable insights with the least technical overhead.

Not all data is used for indicating which features to add to your product. Some data is used to predict a user's wants and needs and then automatically customize features, colors, menus, screens, and notifications (and more) for them. It's personalization on steroids.

Metrics That Matter

In a world where everything can be measured, knowing which metrics truly matter becomes a critical product leadership skill.

North Star Metric

Identifying the single measurement that best captures product success provides clarity and alignment for your entire organization.

Your North Star metric should represent the core value your product delivers to customers. It's not necessarily a business metric like revenue, but rather a measurement of customer value that serves as a leading indicator of business success. The ideal North Star combines three characteristics: it reflects customer value, is sensitive to product changes, and predictively correlates with long-term business outcomes.

Selecting the wrong North Star can be surprisingly damaging, directing team efforts toward outputs that don't actually drive success. I've seen teams chase engagement metrics like "time in app" only to discover they were optimizing for addiction rather than genuine user value. The best North Stars focus on the quality of user experiences rather than just their quantity.

Facebook's shift from total registered users to "monthly active users" and later to "daily active users" illustrates how North Star metrics evolve with product maturity. Early-stage products typically focus on acquisition and activation metrics, while mature products often shift toward retention, engagement quality, or monetization efficiency metrics. Your North Star should evolve as your product and market understanding deepens.

When implementing a North Star metric, ensure everyone understands not just what the metric is, but why it matters and how their work influences it. Break down the North Star into component metrics that different teams can directly impact. Amplitude calls these "input metrics." They're the specific levers teams can pull that collectively drive your North Star. This translation makes the method actionable rather than aspirational.

Leading vs. Lagging Indicators

Distinguishing predictive signals from outcome measurements gives product teams the time to address problems before they impact business results.

Leading indicators predict future performance, while lagging indicators reflect past performance. Revenue and profitability are classic lagging indicators. They tell you how you've done but provide little guidance on future trajectory. User activation rates, engagement patterns, and satisfaction scores typically function as leading indicators that forecast changes in business outcomes.

The challenge is that most organizations overvalue lagging indicators because they're concrete and clearly connected to business goals. Leading indicators often require more interpretation and validation. I've found the most effective approach is to map the causal relationships between your metrics, identifying which early signals reliably precede changes in your core business metrics.

Superhuman, the premium email client, identified a powerful leading indicator in their early development: how many users hit "inbox zero" daily. They found this metric strongly predicted retention and willingness to pay, making it a reliable leading indicator of business health. They focused product development on maximizing this metric, which guided them toward features that genuinely increased user value.

To develop your leading indicators, look for moments of truth in your user journey. These are actions or achievements that separate successful users from those who don't find value. These inflection points often make excellent leading indicators because they signal whether users are on a trajectory toward long-term engagement. For subscription products, this might be a specific usage frequency; for marketplaces, it might be completion of a certain number of transactions.

Engagement Metrics

Tracking how users interact with your product provides vital signals about its relevance and value in their lives.

Engagement metrics measure the frequency, intensity, and quality of user interactions with your product. While simple metrics like daily active users (DAU) or session length provide baseline information, more sophisticated engagement analysis considers interaction patterns and their changes over time.

The most common mistake I see with engagement metrics is focusing exclusively on quantity while ignoring quality. Optimizing for maximum time spent or most features used often leads to cluttered, overwhelming products that ultimately drive users away. Instead, focus on meaningful engagement that delivers user value efficiently.

LinkedIn transformed their product strategy when they shifted from tracking general engagement to monitoring "days members get value," their term for days when users found a job, learned a skill, or made a valuable connection. This reframing focused product development on high-value interactions rather than just increasing time on platform, leading to more satisfied users and, ultimately, better business results.

When analyzing engagement, segment by user type and lifecycle stage. New users engage differently than veterans, and power users have different patterns than occasional users. Looking at average engagement across all users often masks important signals about specific segments' experiences. I recommend creating engagement archetypes based on usage patterns, then tracking how users move between these archetypes over time.

Example: Plantica

Plantica, a houseplant care mobile app, was struggling with user retention despite solid download numbers. The company initially focused on typical engagement metrics like daily active users and time spent in app, which showed steady growth. However, churn rates remained stubbornly high at 73% after the first month.

The turning point came when Plantica's product team shifted from measuring quantity of engagement to quality. Rather than celebrating when users spent more time in the app, they defined "successful plant care events" as their core engagement metric; moments when users followed care advice and their plants thrived as a result.

"We were optimizing for the wrong behavior," admits Jamie Smith, Plantica's founder. "We wanted users to quickly get the information they needed and return to their plants, not endlessly scroll through our app."

The team created engagement archetypes based on usage patterns: casual waterers, plant collectors, and botanical enthusiasts. By tracking how users moved between these archetypes, they identified that plant collectors who successfully care for five or more plants were 4-times more likely to become paid subscribers.

This insight led to a redesigned onboarding flow focused on helping users properly set up multiple plant profiles. Within three months, Plantica saw their paid conversion rate double while average time-in-app actually decreased; proving that for their product, less engagement of higher quality was the path to growth.

Retention Metrics

Measuring ongoing product usage and customer loyalty provides the ultimate verdict on whether you're delivering sustainable value.

Retention metrics track whether users continue engaging with your product over meaningful time periods. While seemingly straightforward, effective retention analysis requires nuance in both measurement and interpretation.

The standard approach to retention — cohort retention curves showing what percentage of users return in subsequent time periods — provides a foundational view but often lacks actionable detail. More sophisticated retention analysis examines retention by user segment, acquisition channel, and feature usage patterns. This multidimensional view helps identify which user types retain best and which aspects of your product drive stickiness.

A common retention analysis mistake is using inappropriate time intervals. Daily retention might make sense for a social media app but would create meaningless noise for a tax preparation product. Match your retention intervals to your product's natural usage frequency. For infrequently used products, consider "return rate when expected" rather than fixed time intervals.

Duolingo revolutionized their approach to retention when they stopped measuring it purely by calendar time and started considering "streak freezes"; a feature allowing users to maintain their learning streak despite missing a day. They discovered users who used freeze features were significantly more likely to return to regular usage rather than churning permanently. This insight helped them design features that accommodated real-life usage patterns rather than demanding perfect consistency.

Decision Architecture

Having data and insights is only valuable if it leads to better decisions. A structured decision architecture transforms analysis into action.

Hypothesis Development

Crafting testable assumptions based on data patterns and user insights forms the bridge between observation and action.

Effective hypotheses have three components: an observation (what you've seen in the data), a proposed explanation (why you think it's happening), and a prediction (what outcome you expect if your explanation is correct). This structure forces clarity of thought and creates a foundation for validation. The format "We believe that [doing X] will result in [outcome Y] because [reason Z]" provides a simple template.

The most common hypothesis development mistake is jumping straight from observation to solution without articulating the underlying belief about user behavior or needs. I've seen teams propose "adding gamification features will increase engagement" without first establishing their hypothesis about why current engagement is low. This shortcut leads to solutions that address symptoms rather than causes.

Amazon's famous six-page memos embody rigorous hypothesis development. Before building new features, teams must articulate the customer problem, their hypotheses about causes and potential solutions, and how they'll measure success. This process forces intellectual honesty and prevents teams from pursuing pet projects without clear expected outcomes.

When facilitating hypothesis development sessions, I've found it valuable to separate the observation phase from the explanation phase. Have the team first document all relevant observations without interpretation, then work together to develop multiple potential explanations for each pattern. This approach prevents premature convergence on a single narrative and encourages considering alternative explanations before committing to solutions.

Confidence Thresholds

Establishing statistical significance levels required before acting on findings helps balance speed with certainty.

Confidence thresholds define how sure you need to be before making a decision. While academics typically require 95% confidence, product teams must balance statistical rigor against the opportunity cost of delay. The appropriate confidence threshold depends on the decision's reversibility, potential impact, and implementation cost.

For easily reversible, low-impact decisions, 80% confidence may be sufficient. For expensive, difficult-to-reverse changes affecting critical metrics, you might require 95% or higher. The key is explicitly defining these thresholds in advance rather than adjusting them to fit desired outcomes — a surprisingly common practice that undermines data-driven decision making.

Booking.com's experimentation culture exemplifies thoughtful confidence threshold implementation. They run hundreds of concurrent experiments, using different significance thresholds based on the test's nature. Minor UI tweaks might require only 80% confidence, while payment flow changes need 99% confidence before implementation. This tiered approach enables rapid iteration where appropriate while maintaining rigor where necessary.

Decision Trees

Mapping potential choices and outcomes creates navigable pathways through complex product decisions.

Decision trees break down complex choices into a series of smaller, more manageable decisions with clear dependencies. They start with a core question, branch into potential answers, and continue branching as each answer raises new questions. This structure forces clarity about decision sequences and contingencies.

The power of decision trees lies in their ability to make assumptions explicit. Each branch represents a specific belief about user behavior or market conditions. This explicitness allows teams to identify which assumptions are most critical and direct research efforts accordingly, rather than trying to validate everything simultaneously.

Intuit's "Follow Me Home" program demonstrates decision trees in action. They observed customers using their tax software, identified key decision points in the user journey, and mapped the various paths users took through the product. This mapping revealed unexpected workflow patterns that informed product improvements. Rather than trying to optimize the entire experience at once, they focused on the most critical decision points first.

When building decision trees, focus on identifying the "load-bearing" assumptions; those that, if wrong, would collapse your entire strategy. These become your research priorities. I've seen teams waste enormous resources validating minor assumptions while leaving critical ones unexamined because they seemed "obvious." Remember that the most dangerous assumptions are often those that seem too basic to question.

Risk Assessment

Evaluating potential downside scenarios when making data-driven product bets helps prevent catastrophic failures.

Risk assessment examines what could go wrong with your decision and creates contingency plans accordingly. The process involves identifying potential failure modes, estimating their likelihood and impact, and developing mitigation strategies for the most serious risks.

The most common risk assessment mistake is considering only the likelihood of success rather than the horrible instance of failure. A 90% chance of success sounds excellent, but if the 10% failure scenario would be catastrophic, the decision deserves more scrutiny. I've found that asking the "reversible" versus "irreversible" question particularly helpful. Irreversible decisions demand much more stringent validation, regardless of the expected upside.

GitHub's gradual rollout strategy works great for risk management. When rolling out significant changes, they deploy to internal users first, then to a small percentage of customers, gradually increasing exposure while monitoring key health indicators. This approach allows them to catch unexpected defects before they affect the entire user base; a strategy that has prevented many potentially damaging releases.

When conducting risk assessments with product teams, I've found it valuable to explicitly designate someone as the "disaster scenario specialist" whose job is to identify everything that could go wrong. This role rotation prevents the common problem of overconfidence bias and ensures thorough consideration of downside scenarios before committing to a direction.

Data Communication

Even the most brilliant analysis creates no value unless it's effectively communicated to decision-makers and stakeholders.

Data Visualization

Creating clear, compelling visual representations transforms complex information into intuitive insights.

Effective data visualization simplifies complexity without oversimplifying. It highlights patterns that matter while filtering out noise. The best visualizations are those where the main insight is immediately apparent even before the viewer reads the title or legend.

The most common data visualization mistake is cramming too much information into a single chart or dashboard. This approach, while well-intentioned, often obscures the key insights. I've found that multiple focused visualizations usually communicate more effectively than a single complex one. Each visualization should answer one specific question clearly rather than vaguely addressing multiple questions.

The New York Times data visualization team demonstrates the power of focused, thoughtful data presentation. Their COVID-19 visualizations made complex epidemiological concepts accessible to the general public by prioritizing clarity over comprehensiveness. They didn't show every possible data point. Instead, they highlighted the patterns most relevant to readers' understanding and decision-making. Let me tell you, I was on that web page everyday during the worst of it; reassured some days, frightened others.

When creating visualizations, start by defining the specific question each chart should answer. Then select the chart type best suited to that question: bar charts for comparisons, line charts for trends over time, scatterplots for relationships between variables, and so on. Finally, eliminate all non-essential elements that don't directly contribute to understanding.

Storytelling With Data

Crafting narratives that connect metrics to meaningful outcomes transforms numbers into compelling calls to action.

Data storytelling combines analytical insights with narrative structure to explain why the data matters and what actions it suggests. Effective data stories have three components: context (what question we're trying to answer), key findings (what the data shows), and implications (what we should do as a result).

The most common data storytelling mistake is presenting insights without connecting them to business or user outcomes. Statements like "conversion increased 15%" remain abstract until framed as "conversion increased 15%, representing 10,000 additional paying customers and $1.2M in annual recurring revenue." This connection transforms data from interesting to actionable. And it's good for early stage companies seeking funding!

Airbnb's data storytelling shows off this principle in action. When presenting findings about their search algorithm improvements, they don't just share the percentage improvements in booking rates. They also translate these into host earnings, nights booked, and market expansion metrics. That way, they connect technical changes to human and business outcomes, making the impact tangible for all stakeholders.

When structuring data stories, I've found the "what, so what, now what" test particularly effective. Start with what the data shows (the objective findings), then explain why those findings matter (the business or user impact), and finally recommend specific actions based on those insights. This progression guides stakeholders from understanding to action rather than leaving them to draw their own conclusions.

Example: CureNext Health

CureNext Health, a telehealth platform connecting patients with specialists, was struggling with a peculiar problem. Their analytics dashboard showed impressive user growth of 27% quarter-over-quarter, yet their customer satisfaction scores were plummeting. The executive team was baffled by these contradictory signals.

Maria Hernandez, their data analyst, recognized that simply presenting these metrics wasn't compelling enough to drive change. Instead of her usual slide deck packed with bar charts, she crafted a data story for the next executive meeting.

She began with a striking visualization: a line graph showing the diverging paths of user growth and satisfaction scores. "This isn't just about numbers trending in opposite directions," Maria explained. "These lines represent approximately 15,000 patients who found specialists but left frustrated with their experience."

Rather than stopping at the what (declining satisfaction despite growth), she moved to the so what (the business impact): "Our data shows that unsatisfied users are 78% less likely to book a follow-up appointment, representing a projected $2.3 million in lost revenue this year alone."

Finally, she presented the now what (the recommended action): "When we segment the data, we discover that satisfaction plummets when patients wait more than 48 hours to hear from specialists. Our matching algorithm is connecting more patients, but not prioritizing response time."

This three-part narrative transformed abstract metrics into a compelling story with clear business implications. The executive team immediately approved resources to redesign the specialist matching algorithm with response time as a key factor. Within two months, satisfaction scores rebounded, and more importantly, follow-up bookings increased by 23%.

"Numbers alone rarely drive decisions," Maria later reflected. "It's when we connect those numbers to patient experiences and business outcomes that data becomes truly powerful."

Cross-Functional Translation

Adapting data insights for different stakeholders enables aligned decision-making across organizational boundaries.

Cross-functional translation means tailoring how you present the same data insights to different audiences based on their priorities, technical background, and decision-making authority. This isn't about changing the substance of your findings, but rather focusing on the aspects most relevant to each audience and using terminology that resonates with their domain.

The most common translation mistake is assuming everyone values the same metrics or speaks the same analytical language. Engineering teams might focus on performance metrics, design teams on usability measures, and executives on growth and revenue indicators. Effective translation connects your insights to each group's primary concerns without losing analytical integrity.

Shopify's product analytics team exemplifies skillful cross-functional translation. When sharing findings about checkout optimization, they translate the same core insights differently for different teams: for designers, they focus on user experience friction points; for engineers, they highlight technical performance issues; for the executive team, they emphasize revenue impact. This tailored approach ensures each group can take appropriate action within their domain.

When preparing cross-functional presentations, I recommend creating a core slide deck with your complete analysis, then developing role-specific versions that highlight the aspects most actionable for each audience. This approach ensures consistent overall findings while maximizing relevance for each stakeholder group.

Insight Prioritization

Distinguishing between interesting findings and actionable intelligence helps teams focus on high-impact opportunities.

Insight prioritization separates signal from noise in your analytical findings. Not all patterns deserve equal attention. Some represent valuable opportunities while others are merely interesting observations. Effective prioritization considers three factors: potential impact, confidence in the finding, and feasibility of action.

The most common prioritization mistake is equating statistical significance with business significance. A highly statistically significant finding with minimal potential impact often receives more attention than a somewhat less certain finding with massive potential upside. I've seen teams chase 1% improvements that were statistically bulletproof while ignoring potential 20% improvements that had more analytical uncertainty.

Google's famous 20% time policy reflects thoughtful insight prioritization at scale. Rather than trying to pursue every promising idea, they created a systematic approach to testing potential opportunities while maintaining focus on core priorities. This balance between exploration and exploitation enabled them to discover breakthrough products like Gmail and Google News without compromising their search business.

When facilitating insight prioritization sessions, use a simple 2x2 matrix of impact versus confidence. High-impact, high-confidence insights become immediate action items. High-impact, low-confidence insights trigger further investigation. Low-impact insights, regardless of confidence, receive lower priority. This method prevents teams from pursuing low-value opportunities simply because they have high analytical certainty.

Conclusion

Mastering data for product decisions isn't about having perfect information; it's about making better decisions with the information you have. The methods, rules, metrics, and approaches we've explored provide a foundation, but the real skill develops through application and reflection.

As you implement these practices, remember that data should inform your decisions, not make them for you. The most successful product leaders combine analytical rigor with experienced judgment, using data to challenge their assumptions while recognizing when to trust their intuition despite analytical uncertainty.

I've seen teams transform their effectiveness by implementing even a few of these practices consistently. Start with the areas most relevant to your current challenges; whether that's improving data quality, refining your analysis methods, focusing on more meaningful metrics, strengthening your decision architecture, or communicating insights more effectively.

The goal isn't becoming a data scientist. It's becoming a product leader who uses data as a powerful tool in service of creating experiences that genuinely improve users' lives. That's the true measure of success — not in the elegance of your dashboards, but in the value your product delivers because of the decisions those dashboards informed.

Concepts in Action: MariTrac

MariTrac, a maritime navigation software startup, began with a promising product that provided real-time weather routing for recreational sailors. Founded by three sailing enthusiasts with tech backgrounds, they launched with enthusiasm but initially struggled to convert their free users to paying customers. Their journey from data confusion to data mastery offers a perfect storm of lessons for product leaders.

Data Integrity

Like many startups, MariTrac's early data collection was haphazard. Their first major product decision — rebuilding their entire notification system — was based on analytics showing users rarely engaged with weather alerts. Three months of engineering work later, they discovered the tracking script had broken during an app update; users had actually been heavily relying on these notifications.

Collection Methods

MariTrac adopted a three-tiered data collection strategy:

1. Core metrics tracking basic product health (daily active users, route completions, subscription conversions)

2. Behavioral data showing how users interacted with specific features

3. Detailed tracking for their most critical feature; storm avoidance routing

This approach prevented the analytics overload that had previously paralyzed their decision-making. More importantly, it revealed a surprising insight: while MariTrac had focused all their development on weather prediction, users primarily valued the social features that allowed them to share their routes with fellow sailors.

North Star Metric

Initially, MariTrac tracked total registered users as their North Star metric. When growth slowed, they realized this wasn't reflecting actual product success. After analyzing their data, they identified "Routes Completed Safely" as their new North Star; a metric that combined completion rate with weather condition severity.

"What makes this North Star powerful is that it directly reflects the value we provide to customers," explains a team member. "When we optimize for routes completed safely, we're focusing on the outcome our users actually care about, not just engagement for engagement's sake."

Leading vs. Lagging Indicators

MariTrac discovered that "weather data consultation frequency" was a powerful leading indicator of retention. Users who checked weather updates at least three times before a sailing trip were 78% more likely to remain subscribers than those who didn't, regardless of whether they ultimately used the route recommendations.

This insight transformed their onboarding process. Rather than pushing new users to complete routes, they encouraged weather data exploration through targeted tutorials. Retention improved by 23% within one quarter of implementing this change.

Hypothesis Development

When faced with lower-than-expected conversion rates from free to paid plans, MariTrac developed a clear hypothesis: "We believe that free users don't convert because they don't experience our premium routing features in challenging weather conditions. If we offer limited access to premium features during forecasted storms, conversion rates will increase by at least 15%."

This structured hypothesis led to a successful limited-time feature unlock strategy tied to weather forecasts, which exceeded their conversion targets by 7%.

Data Communication

The company transformed their monthly analytics reviews by tailoring presentations to different stakeholders. For engineers, they focused on system performance during peak weather events. For marketers, they highlighted acquisition channels with the highest retention rates. For executives, they translated all metrics into revenue impact and market penetration.

"The real breakthrough came when we stopped treating data as just numbers and started turning it into stories," says the CEO. "We'd show how a specific sailor avoided a dangerous storm system using our premium features, then connect that personal story to our broader metrics about routes completed safely."

Results

Three years after their data transformation began, MariTrac achieved profitability with a 42% year-over-year growth rate. Their subscription retention reached 89%; well above industry averages. Perhaps most tellingly, they reduced development cycles by 50% by eliminating features that data showed wouldn't impact their North Star metric.

"The biggest lesson we learned," reflects a team member, "is that data mastery isn't about having perfect information. It's about making better decisions with the information you have, while knowing which questions are most important to answer next."

Execution Essentials

Data Collection & Analysis Methods

- Always verify data integrity before making major decisions; a broken tracking script can waste months of development effort.

- Implement a tiered data collection strategy focusing first on core metrics, then expand to more granular behavior tracking.

- Use cohort analysis to separate users into groups based on shared characteristics to reveal patterns that aggregate metrics often mask.

- Apply funnel analytics to identify exactly where users abandon your critical flows and focus optimization on high drop-off points.

- When A/B testing, define clear, predetermined success metrics before launching any test.

Metrics That Matter

- Select a North Star Metric that reflects customer value, is sensitive to product changes, and correlates with business outcomes.

- Evolve your North Star as your product matures; early-stage products typically focus on acquisition while mature products shift toward retention.

- Identify leading indicators that predict future performance rather than just measuring past results.

- Look for "moments of truth" in your user journey that separate successful users from those who don't find value.

- Focus on meaningful engagement that delivers user value efficiently rather than just maximizing time spent or features used.

Decision Architecture & Data Communication

- Structure hypotheses with three components: an observation, a proposed explanation, and a prediction.

- Set different confidence thresholds based on a decision's reversibility, potential impact, and implementation cost.

- Create focused visualizations where each chart answers one specific question clearly rather than addressing multiple questions vaguely.

- Use the "what, so what, now what" test to structure data stories that guide stakeholders from understanding to action.

- Tailor how you present the same data insights to different audiences based on their priorities and domain knowledge.

Implementation Strategies

- Start with the areas most relevant to your current challenges rather than trying to implement all practices at once.

- Create separate "ledgers" for qualitative and quantitative insights, then cross-reference them to identify alignment and divergence.

- For each metric you track, document exactly what it represents and how it's calculated to prevent misunderstandings.

- Gradually roll out significant changes to small percentages of users while monitoring key health metrics before full deployment.

- Remember that data should inform your decisions, not make them for you; combine analytical rigor with experienced judgment.

The Product
Chapter 15

Extensibility, Partnerships & Integrations

I've learned that truly great products rarely exist in isolation. The most successful products today don't function as islands. They're more like continents with well-established trade routes, shared resources, and mutual dependencies. This chapter explores how to think beyond your product's current boundaries through extensibility, partnerships, and platform integrations.

When I was young, I built products with clearly defined edges. We controlled every pixel, every interaction, every data point. Today's product leaders face a different reality: your users expect your product to work seamlessly with their existing tools and workflows. They expect customization options. They expect your product to grow and evolve with their needs. And in many cases, they expect you to let others build on top of your foundation.

What follows is a practical guide for product leaders navigating these interconnected waters. We'll explore strategies for making extensible products, forming strategic partnerships, and integrating with platforms your users already rely on; all while maintaining focus on your core value proposition and not spreading your team too thin. Because let's face it, in product management, we're always balancing the excitement of possibility with the constraints of reality.

Product Extensibility

The ability to extend your product beyond its initial scope is crucial for long-term success. In this section, we'll examine how to build extensibility into your product from the ground up.

When I think about product extensibility, I'm reminded of conversations with a PM on my team who once asked, "Why are we spending so much time building something nobody's asking for yet?" He was referring to our API design process, which was taking valuable engineering time. Six months later, when our biggest customer requested a custom integration that would have taken months without that API groundwork, he understood. Extensibility isn't just about preparing for the unknown; it's about creating options for your product's future.

Building Modular Architecture

A modular architecture creates the foundation upon which all your extensibility efforts will stand or fall. Think of it as the difference between a solid brick wall and a set of well-designed building blocks.

Modularity begins with your technical architecture, but it's fundamentally a product decision that requires close collaboration between product and engineering leadership. When designed well, modular systems allow you to swap, extend, or enhance components without rebuilding the entire structure. This is particularly valuable when market demands shift unexpectedly, as they inevitably do.

The key to successful modular design is identifying the right level of abstraction. Too granular, and your team drowns in complexity; too coarse, and you lose the flexibility benefits. I've found that starting with major functional domains as modules, then progressively refining them as you understand usage patterns, strikes the right balance. For example, in a marketing automation product, you might separate campaign management, analytics, and content creation into distinct modules that can evolve independently.

One often-overlooked aspect of modularity is establishing clear contracts between modules. These contracts — typically in the form of well-defined interfaces — ensure that changes to one module don't unexpectedly break another. Slack did this exceptionally well; their message composition, delivery, and display systems work as separate modules with clearly defined interactions, allowing them to progressively enhance each without disrupting the others.

Remember that modularity comes with costs. Development might initially take longer, and your architecture will be more complex. The ROI comes later when changes that would have taken months take days instead. Google's Android team exemplifies this approach. They designed their permission system as a modular component, which allowed them to completely overhaul it in Android 6.0 without rebuilding the entire operating system.

Designing Flexible APIs

Application Programming Interfaces (APIs) serve as the bridges connecting your product to the outside world. Flexible, well-designed APIs open possibilities far beyond what your team can build alone.

Imagine you're the Weather Channel, and you're all about giving people the latest weather updates. Why not make your forecasts available through an API? That way, other websites and apps can use your data and create their own weather displays. And guess what? You can charge a fee for the information, while the apps can design their own interfaces.

API data is hidden from the user. It's usually sent by web servers. When an app needs data, it makes a request to the server (through a URL) in the background, gets the data, assembles it, and then shows it to the user in your app's interface. The user gets more info and it's all smooth sailing.

When designing APIs, prioritize consistency, simplicity, and forward compatibility. Consistency means following established patterns, whether RESTful principles, GraphQL schemas, or another standard. This reduces the learning curve for developers using your API. Simplicity involves exposing only what's necessary to accomplish the task at hand, without leaking implementation details. Forward compatibility means designing with future versions in mind, so you don't break existing integrations when you iterate.

I've learned that truly developer-friendly APIs are built with specific use cases in mind, not theoretical abstractions. Start by identifying the top three integration scenarios your customers might want, then design APIs that make those scenarios delightfully simple. Stripe exemplifies this with their payment processing API. It's designed specifically to make accepting payments as easy as possible, not to expose every internal system.

Coming back to our weather example, if you're the weather forecast provider, you would figure out your most important API use cases (external customers). One I can think of right away are operating systems. Think Windows, Mac, and Linux. They have widgets for this type of heads-up information.

Once you have your API ready to go, consider that documentation quality often separates successful APIs from failed ones. Technical writers are worth their weight in gold here. Your documentation should include clear explanations, practical examples, and ideally, interactive tools to help developers test API calls without writing full integrations. When Twilio launched, their documentation included working code examples in six programming languages, dramatically lowering the barrier to adoption.

Don't forget that APIs have unique security considerations. Rate limiting, authentication, data validation, and access controls are non-negotiable components. The best API designs build these elements in from the beginning rather than retroactively. The financial cost of an API security issue (breach) can be astronomical, as Venmo discovered when early versions of their API exposed transaction histories that users thought were private.

Creating Developer-Friendly Documentation and SDKs

Great documentation transforms your extensibility features from theoretical capabilities to practical tools. It's the difference between having extensibility and having extensions.

Developer documentation requires a unique approach compared to end-user help guides. Developers need both conceptual overviews that explain your system's architecture and practical references that detail every parameter, error code, and response format. Stripe's documentation excels at addressing both needs. They provide an architectural overview that explains their resource model while also offering detailed API references with live code examples.

Software Development Kits (SDKs) take your API a step further by packaging common interactions into language-specific libraries. Well-designed SDKs handle the boilerplate code for authentication, error handling, and data formatting, allowing developers to focus on their unique integration needs. For example, while GitHub's API is comprehensive, their SDKs for popular languages like Python, JavaScript, and Ruby make it significantly easier for developers to interact with repositories programmatically.

The best documentation includes more than just technical details. It also includes onboarding tutorials, conceptual guides, and sample applications. Twilio provides complete sample applications showing how to build common use cases like two-factor authentication or automated notifications, dramatically reducing the time to first success for new developers.

Remember that documentation and SDKs aren't one-time investments. They require maintenance alongside your product. When Square updated their point-of-sale API, they simultaneously updated their documentation, samples, and SDKs across seven programming languages. This commitment to keeping extension tools current directly correlates with adoption rates among third-party developers.

Example: StormSight

StormSight began as a modest weather forecasting website serving rural Midwest counties, founded by two meteorologists and a developer named Michael. After attracting 15,000 monthly visitors with their accurate local forecasts, they noticed an interesting pattern in user requests.

Farmers wanted rainfall predictions in their crop management software. Event planners asked to embed forecast widgets on websites. Even the county emergency services inquired about integrating storm alerts with their notification system.

Michael recognized the opportunity. "We're sitting on valuable data that people want to use in different contexts," he explained. "Instead of building one-off solutions, we should make our system extensible."

Despite initial skepticism from the meteorologists who worried about diverting resources from forecast improvements, Michael convinced them to undertake a six-week refactoring project. They separated their system into distinct modules: data collection, forecast generation, alert systems, and user interfaces — each communicating through clearly defined interfaces.

This modularity immediately improved their development process. The team could update individual components without breaking others. Building on this foundation, Michael designed a REST API exposing key functionality to external developers, focusing on common scenarios: accessing current conditions, retrieving forecasts, and setting up custom alerts.

The turning point came when they hired Jack, a technical writer who created comprehensive, engaging documentation. The team also built SDKs for JavaScript, Python, and PHP that handled authentication, error handling, and data formatting — reducing integration time from days to hours.

Six months later, the results exceeded expectations. A farming platform integrated StormSight's precipitation forecasts into their planting recommendations. A local developer built a smart home integration that closed windows when rain was imminent. The county emergency services created a notification system combining StormSight's data with evacuation protocols.

Within two years, 60% of their data requests came through the API. They introduced tiered pricing for commercial users while keeping basic access free for hobbyists and educational projects, creating a new revenue stream that funded additional meteorologists.

As Michael now says during their developer meetups, "We stopped trying to build everything ourselves and started enabling others to build with us. We provide the data, you provide the context that makes it relevant."

Implementing Plugin Systems

Plugin architectures represent the highest form of product extensibility, allowing third parties to extend your product's functionality without your direct involvement in every enhancement.

Successful plugin systems define clear boundaries between core product functionality and extensible areas. WordPress exemplifies this approach, with a stable core that handles content management fundamentals while allowing plugins to extend nearly every other aspect of the platform. This separation ensures the core product remains stable while enabling incredible customization freedom.

The technical implementation of your plugin system creates inherent trade-offs between security, performance, and flexibility. More powerful plugin capabilities increase security and stability risks. Chrome's extension system illustrates a thoughtful approach to this balance. Extensions run in sandboxed environments with explicitly declared permissions, limiting potential damage while still enabling a lot of browser enhancements.

Documentation is even more critical for plugin systems than for APIs. Beyond explaining technical interfaces, you need to establish design guidelines, review processes, and best practices. These resources help create consistency across the ecosystem and reduce support burdens on your team. Salesforce's AppExchange includes comprehensive guidelines for security, user experience, and performance that all third-party developers must follow.

Don't underestimate the community-building aspect of plugin ecosystems. Active developer forums, showcase opportunities, and recognition programs create incentives beyond financial ones. Figma's plugin community thrives partly because they highlight innovative plugins in their communications and provide spaces for plugin developers to connect with each other.

Balancing Customization with Simplicity

The paradox of extensibility is that too many options can reduce usability for your core audience while too few limit your product's growth potential.

Finding this balance begins with understanding your user segments and their technical sophistication. Enterprise customers typically need extensive customization options, while mainstream users often prefer simplicity with thoughtful defaults. Products like Notion navigate this balance well. They offer a simple to use interface for basic users while also providing powerful customization features for those who need and want them.

Progressive disclosure of extensibility features helps manage complexity. Start with an interface that works well for common use cases, then provide clear paths to customization for users who need it. Airtable takes this approach by offering templates for common workflows while making advanced features like API access, automations, and custom applications accessible but not overwhelming.

Maintain strong defaults alongside customization options. Well-designed defaults ensure that basic users get a solid experience without any customization, while advanced users can tailor the product to their specific needs. Shopify's theme system exemplifies this approach. Their default themes work beautifully out of the box, but can be customized extensively for unique brand experiences.

Remember that every customization option creates testing complexity and potential support burden. Be intentional about which aspects of your product you make customizable, focusing on areas where user needs genuinely diverge. Apple's iOS (phone and tablet operating systems) demonstrates this philosophy. While they've continuously added customization options over the years, they maintain fairly tight control over core system behaviors in order to ensure consistent quality.

Strategic Partnerships

No product exists in a vacuum. Strategic partnerships allow you to leverage others' strengths while focusing on your own core competencies.

Partnerships have dramatically evolved from the old-school business development deals I witnessed early in my career. Those were often more about press releases than user value. Today's most effective product partnerships are deeply integrated experiences that feel natural to users. They're less about logos on websites and more about creating seamless workflows across product boundaries.

Identifying Complementary Products

Finding the right partners begins with mapping your users' workflows beyond your product's boundaries. Where do they go before using your product? Where do they go after?

The most valuable partnerships often emerge at these workflow transition points. Look for products that are adjacent to yours without directly competing. Slack and Zoom demonstrate this principle well. They serve related but distinct communication needs (asynchronous text vs. synchronous video), creating a natural partnership opportunity that enhances both products' value propositions.

Data can also reveal partnership opportunities you might not intuitively recognize. Analyze how users integrate your product with others through whatever means available: authorized API connections, manual imports/exports, or even survey responses. HubSpot discovered through usage analysis that their marketing customers were frequently copying data to accounting systems, leading to their partnership and integration with QuickBooks.

Consider partnerships that bring capabilities outside your core focus. Building everything yourself leads to mediocrity across many features rather than excellence in your unique value proposition. When Figma partnered with various prototyping tools rather than building comprehensive prototyping capabilities internally, they maintained their focus on being the best collaborative design tool while still meeting their users' full workflow needs.

Don't overlook partners based on size mismatches. A smaller company might be more motivated to create a deeply integrated experience with your product, while larger companies might offer greater market reach. DocuSign's early partnership with Salesforce, when DocuSign was much smaller, created substantial value for both companies despite their size difference.

So, here's something important to consider. Don't get too carried away with loving a partner product. Unless you're absolutely certain the company is financially stable, it's best to keep your distance. Make sure their core product is compatible with yours and will remain compatible. The last thing you want, once you've got millions of users, is to suddenly lose features because of a partner's financial troubles, merger, or just because they decide to stop supporting you. I will also talk about managing partner life-cycles later.

Developing Partnership Evaluations

Not all potential partnerships deserve your investment. A systematic evaluation process helps identify the ones worth pursuing.

Start with strategic alignment between your companies. This goes beyond market fit to include compatible visions, cultural alignment, and compatible timelines. The partnership between Spotify and Uber worked because both companies prioritized creating memorable user experiences and were willing to invest in unique integrations rather than standard API connections.

Quantify the potential user value of the partnership. This should include both acquisition potential (new users you might gain) and retention impact (existing users who will find more value). Quantified value estimates create the basis for fair value exchange discussions later. When Microsoft and Adobe partnered to integrate Creative Cloud with Office 365, both companies had clear metrics around how the integration would benefit their respective users.

Assess technical compatibility and integration complexity honestly. A partnership that would require months of engineering work needs to deliver proportionally higher value than one requiring minimal technical investment. The Netflix and cable provider partnerships succeeded partly because Netflix invested heavily in reference designs that made integration technically straightforward for providers.

Don't neglect the human factor. Partnerships require ongoing collaboration, so evaluate the working relationship with your potential partner during the exploration phase. Are they responsive? Do they follow through on commitments? Do your teams work well together? These factors often determine success more than the technical or business aspects of the partnership.

Creating Mutual Value Exchange Models

Sustainable partnerships deliver clear value to both companies and their users. One-sided partnerships rarely last.

The foundation of mutual value is understanding exactly what each party needs from the relationship. Sometimes it's revenue, sometimes it's distribution, sometimes it's enhanced product capabilities. Being explicit about these needs early prevents misaligned expectations later. The Apple and Nike partnership works because Apple gains specialized fitness content and hardware differentiation, while Nike extends their brand relationship with customers into the digital realm.

Formalize the value exchange in measurable terms. Rather than vague benefits, define specific KPIs each partner aims to achieve. This might include new user acquisition targets, engagement metrics for integrated features, or revenue goals for joint offerings. Airbnb and American Express established clear goals for their partnership: Amex wanted increased card usage for travel, while Airbnb wanted access to Amex's premium customer base.

Create incentive structures that align behavior with desired outcomes. If your goal is increasing adoption of the integrated experience, design partnership terms that reward that specific outcome rather than just logo placement or marketing mentions. The Uber and Spotify integration succeeded because both companies prioritized and incentivized actual usage of the feature, not just its existence.

Be prepared to evolve the value exchange over time as market conditions and product priorities change. The most durable partnerships have built-in mechanisms for reassessing and adjusting the relationship. Facebook and Shopify have maintained a productive partnership through numerous platform changes because they regularly reassess how they can create mutual value as e-commerce and social media landscapes evolve.

Establishing Governance for Partner Integrations

Clear governance processes prevent partnership chaos while maintaining quality standards for your users.

Start by defining decision rights; who can approve what, and where consensus is required versus where one partner has final say. Generally, each company should maintain control over their core user experience while establishing collaborative processes for the integration points. Salesforce's AppExchange governance model clearly delineates where Salesforce maintains control and where partners have autonomy.

Quality standards protect both parties' brands. Establish clear criteria for performance, reliability, security, and user experience that integrated experiences must meet. Microsoft's partnership program includes detailed technical requirements that partners must satisfy before being certified, protecting both Microsoft's reputation and the partner's investment.

Create communication protocols for both routine updates and crisis situations. When an issue inevitably arises, having established channels and response expectations prevents minor problems from becoming partnership-threatening events.

Don't neglect user support governance. Define how support requests related to the integration will be routed, who is responsible for fixing various types of issues, and what the expected resolution timeframes are. The partnership between Dropbox and Microsoft Office includes clear procedures for determining whether an issue stems from Office, Dropbox, or the integration between them.

Managing Partnership Lifecycles

Like products, partnerships have natural lifecycles from initial exploration through maturity and eventually to sunset.

The exploration phase should focus on validating the strategic and technical fit before significant resources are committed. Small proof-of-concept projects help both parties evaluate the working relationship and integration complexity with minimal investment. Before fully committing to their partnership, Lyft and Google Maps created a limited integration to assess user response and technical compatibility.

Launch planning requires close coordination of technical rollouts, marketing activities, and support training. The most successful partnership launches include joint testing periods, coordinated communications, and prepared support teams on both sides. When Peloton and Spotify launched their integration, they coordinated announcements across multiple channels and prepared detailed support documentation for the inevitable questions.

Maintenance is where many partnerships falter without proper attention. Establish regular check-ins to review performance metrics, discuss upcoming product changes that might affect the integration, and identify enhancement opportunities. Amazon and Ring maintain their successful partnership through quarterly business reviews where they assess integration performance and plan future enhancements.

Have honest conversations about potential sunset scenarios before they arise. Define the conditions under which the partnership would end and establish orderly wind-down processes that minimize user disruption. When IFTTT and Nest ended their partnership, they implemented a three-month transition period with clear user communication, preserving goodwill despite the discontinuation.

Example: Chow Magic

Chow Magic started as a simple meal planning app that helped busy professionals plan their weekly meals. Their core product was solid — users loved the recipe suggestions and automatic grocery lists — but CEO Patty noticed something interesting in their user data: most of their customers were manually copying shopping lists into grocery delivery apps.

"We were making half the process easier, but then forcing users to do this annoying copy-paste dance," Patty explained. "Our team was small, and building our own grocery delivery network would've been insane."

Instead of trying to become a delivery company, Patty identified Speedcart — a growing grocery delivery service — as a complementary product. Their services were adjacent but non-competing, creating what the partnership team called a "workflow transition point" where users naturally moved from one product to the other.

Rather than just slapping logos on each other's websites, they created a deeply integrated experience. Users could plan meals in Chow Magic and then with one tap, send their grocery list directly to Speedcart for delivery scheduling.

The partnership evaluation process was surprisingly systematic. Both companies quantified potential value — Chow Magic estimated a 15% retention improvement, while Speedcart projected 10,000 new users in the first quarter. They also assessed technical compatibility before committing resources.

"The human factor was huge," Patty recalled. "We clicked with their product team right away. When integration challenges came up, nobody pointed fingers — we just fixed stuff together."

The partnership agreement formalized clear KPIs for both sides, with incentives tied to actual usage of the integrated feature, not just marketing mentions. They also established governance protocols for quality standards and support — when users had problems, both teams had clear guidelines about who handled what.

Three years in, they still conduct quarterly reviews to assess performance and plan enhancements. Recently, they expanded the integration to include recipe suggestions based on sale items at local Speedcart partner stores — something neither company could have done alone.

"The best partnerships solve real user problems while letting each company stick to what they do best," Patty says. "We didn't try to become delivery experts, and they didn't try to become meal planning gurus. We just built a bridge between our islands."

Platform Integration

Beyond direct partnerships, integrating with established platforms can dramatically extend your product's reach and capabilities.

I remember a pivotal strategy meeting where we were debating building our own proprietary system versus integrating with an established platform. A senior engineer made a comment that has stuck with me: "We can build a better mousetrap, or we can plug into where the mice already live." That pragmatic insight guided our decision to integrate rather than reinvent, saving months of development and instantly connecting us with our target users.

Analyzing User Platform Dependencies

Effective platform strategy starts with understanding which platforms matter most to your specific users. Not all platforms are equally relevant to your product.

Begin by mapping your users' existing tool ecosystems. What platforms do they already rely on daily? Where do they store critical data? What devices do they use? This mapping should draw on both quantitative data from surveys or usage analytics and qualitative insights from user interviews. When Notion analyzed their users' workflows, they discovered Slack was a critical communication hub for their target audience, leading to their prioritization of rich Slack integration features.

Analyze the stability and trajectory of potential platform partners. Integrating with declining platforms can waste resources, while missing emerging platforms can create strategic disadvantages. Zoom quickly recognized Microsoft Teams' growing enterprise adoption and prioritized deep Teams integration despite Teams including competing functionality.

Consider platform economics alongside user needs. Some platforms charge prohibitive fees or impose restrictive terms that may not justify the integration investment. Others offer revenue-sharing or co-marketing opportunities that improve your ROI calculation. Strava, a fitness tracking app, carefully evaluates the economic terms of each device platform they integrate with, prioritizing those with favorable data-sharing terms and large active user bases.

Don't ignore geographic differences in platform dominance. WhatsApp might be essential in Brazil but less critical in Japan. WeChat integration might be irrelevant for a US-focused product but mandatory for success in China. Spotify's platform integration strategy varies significantly by region, with deeper integration into platforms that dominate specific markets.

Building Seamless Authentication Experiences

Authentication represents the first interaction touchpoint in any platform integration, setting expectations for the entire experience.

Single Sign-On (SSO) capabilities reduce friction and increase adoption rates for integrated experiences. When implemented well, users should barely notice the transition between your product and the platform. The integration between Asana and Google Workspace demonstrates this principle. Users can access Asana features directly within Google applications using their existing Google account credentials.

Design for the authentication flows with the greatest security and the least friction. OAuth flows have become standard for good reason, balancing security needs with user experience. Trello's implementation of Google and Microsoft authentication follows best practices by limiting permission requests to only what's absolutely necessary and clearly explaining why those permissions are needed.

Consider the enterprise authentication requirements of your target customers. Large organizations often require SAML, JIT provisioning, and specific security certifications. Box prioritized enterprise authentication standards early, making them the preferred content platform for security-conscious organizations despite competing with more consumer-friendly options.

Remember that authentication is an ongoing relationship, not a one-time event. Design for smooth re-authentication experiences, expired token handling, and permission updates. Slack's Google Drive integration handles authentication token refreshes invisibly to users, maintaining the connection without interrupting workflows.

Designing Data Synchronization Methods

Keeping data consistent across integrated systems presents complex technical and user experience challenges.

Start by determining the appropriate synchronization model: real-time, scheduled, or manual. This decision should balance technical constraints with user expectations. Real-time sync creates the most seamless experience but requires more robust infrastructure. When Monday.com integrated with Jira, they implemented near-real-time synchronization for critical fields while using scheduled syncs for less time-sensitive data, striking a practical balance.

Create clear conflict resolution protocols. When the same data is modified in multiple systems, how do you determine which change prevails? The best integrations make these rules transparent to users and provide mechanisms for resolving conflicts when automatic resolution isn't possible. Dropbox's Microsoft Office integration includes visual indicators showing sync status and conflict notifications that explain the issue and offer resolution options.

Consider data volume and performance implications, especially for mobile users. Overly aggressive synchronization can drain batteries and consume excessive bandwidth. Evernote's platform integrations use intelligent syncing algorithms that prioritize recent and frequently accessed notes while deferring sync for less relevant content when on cellular connections.

Don't overlook privacy implications of data synchronization. Users should maintain control over what information flows between systems, with clear explanations of data sharing terms. LastPass's browser extensions explicitly request permission before synchronizing data with new devices and provide granular controls over which information is shared.

Creating Fallback Mechanisms

Even the most reliable platforms experience downtime. Building resilience into your integrations preserves user trust during inevitable disruptions.

Design graceful degradation patterns that maintain core functionality when integrated platforms are unavailable. This might mean caching critical data locally or providing alternative workflows. Todoist's Gmail integration continues to display previously synced tasks even when Gmail's API is unreachable, allowing users to reference their task list during Gmail outages.

Implement proactive monitoring to detect integration issues before users report them. This includes watching for increased error rates, performance degradation, or functionality changes in platform APIs. When Zapier detects issues with a platform API, they automatically display status notifications to affected users rather than waiting for users to discover the problem.

Create transparent communication mechanisms that keep users informed about integration status. When issues occur, users should understand what functionality is affected and what alternatives are available. Zendesk's status page breaks down the operational status of each integrated platform, allowing customers to quickly determine whether an issue stems from Zendesk or an integrated service.

Test fallback mechanisms regularly, not just during initial development. Platforms change, and without regular testing, your fallback systems might silently fail when needed most. Netflix's chaos engineering approach includes regularly simulating the failure of integrated authentication and payment systems to ensure their fallback mechanisms function correctly.

Measuring Integration ROI

Not all integrations deliver equal value. Systematic measurement helps you double down on successful integrations and reconsider underperforming ones.

Establish baseline metrics before launching platform integrations. These should include both direct integration metrics (adoption rates, usage frequency) and broader product impacts (retention changes, feature usage patterns). When Hubspot launched their Salesforce integration, they measured not only integration usage but also how it affected overall platform engagement and customer retention rates.

Create attribution models that correctly identify the impact of platform integrations on user acquisition and retention. Multi-touch attribution models often provide more accurate insights than simplistic first-touch or last-touch approaches. Buffer attributes new user acquisition to specific platform integrations by tracking the complete user journey from initial touchpoint through conversion.

Analyze integration usage patterns to identify enhancement opportunities. Are users leveraging all integration capabilities or just a subset? Are there common points where users abandon integrated workflows? Mailchimp uses these insights to regularly refine their Shopify integration, focusing development resources on the integration points that deliver the most user value.

Remember that ROI calculations should include maintenance costs, not just initial development investments. Some integrations require significant ongoing resources to maintain, reducing their long-term value. Tableau's platform integration strategy includes regular ROI reassessments that factor in both ongoing engineering costs and customer value delivered, leading to occasional strategic decisions to sunset less valuable integrations.

Conclusion

As I reflect on the evolving landscape of product extensibility, partnerships, and platform integrations, one thing becomes increasingly clear: the role of product leadership itself is changing.

Today's most effective product leaders think beyond their product's boundaries. They see their offering not as an isolated solution but as a node in an interconnected ecosystem. This perspective shift changes how we build products, how we measure success, and how we create value for our users and businesses.

The skills required for this connected product leadership approach are multidisciplinary. Technical understanding helps you evaluate architectural decisions that impact extensibility. Business acumen guides partnership evaluations and platform strategy. Communication skills become even more critical when coordinating across organizational boundaries. And a user-centered mindset ensures that all these connections actually enhance rather than complicate the user experience.

As you implement the strategies outlined in this chapter, remember that the goal isn't connectivity for its own sake. The goal remains creating exceptional user experiences and business value. Connectivity is simply a powerful means to that end. The most successful connected products maintain a clear core value proposition while thoughtfully extending their reach through the approaches we've explored.

If I could leave you with one final thought, it's this: building connected products requires both vision and pragmatism. The vision to see beyond your product's current boundaries and imagine new possibilities through extensibility, partnerships, and platform integrations. And the pragmatism to make disciplined choices about which connections truly matter to your users and merit your limited resources.

Concepts in Action: TractorTech

TractorTech is an e-commerce platform that specializes in heavy equipment parts for construction, mining, and agricultural industries. What began as a simple on-line catalog has evolved into a comprehensive platform that exemplifies the extensibility concepts outlined in the chapter.

Building Modular Architecture

From day one, TractorTech designed their platform with modularity in mind. The website is structured around four core modules: parts inventory, equipment compatibility, order management, and technical documentation. Each module functions independently yet communicates seamlessly with the others through well-defined interfaces.

This modular approach paid dividends when supply chain disruptions occurred during global economic shifts. While competitors struggled to adapt their systems, TractorTech simply enhanced their inventory management module to incorporate alternative suppliers, without disrupting the rest of the system. This flexibility allowed them to respond to market changes in weeks rather than months.

API Development and Documentation

The company recognized early that their customers used numerous specialized tools for fleet management, maintenance scheduling, and procurement. Instead of trying to build everything themselves, they created a flexible API that allows their platform to integrate with maintenance systems, inventory management software, and enterprise resource planning platforms.

Their API documentation is exceptional. It includes conceptual overviews that explain the system architecture, detailed reference guides for every endpoint, and interactive examples that let developers test API calls directly in the browser. For popular programming languages, TractorTech provides SDKs that handle authentication, error handling, and data formatting, making integration significantly easier.

Plugin Ecosystem

Taking extensibility further, TractorTech implemented a plugin system that allows third-party developers to extend the platform's functionality and create white-labeled, fully customizable websites. Their plugin architecture maintains a careful balance between power and security. Plugins run in sandboxed environments with explicitly declared permissions, limiting potential security risks while enabling significant enhancements.

The company cultivates an active developer community through regular webinars, showcase opportunities, and a recognition program that highlights innovative plugins. This community-building effort has resulted in over 200 plugins that address specialized needs TractorTech couldn't prioritize internally, such as custom reporting for specific equipment types and specialized warranty tracking tools.

Strategic Partnerships

TractorTech has formed strategic partnerships with complementary products that serve their heavy equipment customers. Their most successful partnership is with FleetTrack, a maintenance management platform.

Before pursuing the partnership, the company conducted a thorough evaluation. They verified strategic alignment between the companies, quantified the potential user value, and assessed technical compatibility. Most importantly, they established a clear mutual value exchange model: TractorTech gained enhanced maintenance recommendation capabilities, while FleetTrack accessed a streamlined parts procurement process.

The partnership governance includes defined decision rights, quality standards for the integrated experience, and clear support protocols. Quarterly business reviews keep both companies aligned as their products evolve.

Platform Integration

Beyond direct partnerships, TractorTech integrates with platforms their customers already use. After analyzing user dependencies, they prioritized integrations with major ERP systems like SAP and Oracle, as well as specialized construction management software.

Their authentication implementations create seamless experiences through Single Sign-On, requiring minimal steps for users to connect their accounts. Data synchronization between platforms is handled thoughtfully, with appropriate models for different data types. Time-sensitive information like inventory levels syncs in real-time, while less critical data like historical purchase records syncs on a scheduled basis.

Understanding that even reliable platforms experience downtime, TractorTech built robust fallback mechanisms. When an integrated platform is unavailable, their website gracefully degrades by caching critical data locally and providing alternative workflows. Their proactive monitoring system detects integration issues before users report them, maintaining trust through transparency.

Measuring Success

TractorTech maintains discipline in their extensibility efforts by rigorously measuring ROI. Before launching any integration, they establish baseline metrics for both direct integration usage and broader business impacts. Their sophisticated attribution model correctly identifies how platform integrations affect customer acquisition and retention.

This measurement discipline has led them to sunset underperforming integrations while doubling down on those delivering the most value. For example, after discovering that their integration with a smaller fleet management system had low adoption rates despite high maintenance costs, they redirected those resources toward enhancing their integration with industry-leading platforms, which showed stronger usage patterns and positive retention impacts.

TractorTech exemplifies the connected product leadership approach described in this chapter. They maintain a clear core value proposition while thoughtfully extending their reach through extensibility, partnerships, and platform integrations. Their success demonstrates that in today's interconnected product landscape, the most valuable solutions are those that work seamlessly with the broader ecosystem rather than trying to exist as islands.

Execution Essentials

Product Architecture and Design

- Design modular systems with well-defined interfaces to enable independent evolution.

- Create APIs focused on top customer integration needs with consistent patterns.

- Develop SDKs with comprehensive documentation covering concepts and details.

- Implement sandboxed plugin systems with clear boundaries around core functionality.

Strategic Partnerships

- Identify partners by mapping user workflows and verifying financial stability.

- Evaluate alignment, quantify value, and assess technical compatibility before committing.

- Define measurable KPIs and incentives that reward usage, not just placement.

- Establish clear governance including decision rights, quality standards, and support roles.

- Test relationships with small projects first and plan for eventual sunset scenarios.

Platform Integration Strategies

- Focus on platforms your users already depend on, considering regional differences.

- Create frictionless authentication with minimal permissions and invisible token refreshes.

- Select appropriate sync models based on data criticality and performance needs.

- Build graceful degradation pathways for when connected platforms fail.

- Measure both direct integration usage and broader product impact metrics.

Building Connected Products

- Start with major modules then refine as usage patterns emerge.

- Create interactive API documentation and update all materials simultaneously when changing.

- Cultivate developer communities through recognition and show-case opportunities.
- Coordinate cross-company launches with aligned technical, marketing, and support teams.
- Regularly test integration failure scenarios to verify fallback mechanisms.

Measuring Success and Evolution

- Implement a tiered approach to help you through interface complexity.
- Adopt detailed technical requirements for partnership quality.
- Implement varied sync frequencies based on data importance like Monday.com.
- Maintain invisible authentication refreshes for seamless connections.
- Create regular ROI assessments to guide sunsetting decisions.

The Product

Chapter 16

Go-To-Market Planning

The moment of truth for any product isn't when the final line of code is written or when the design is perfected; it's when it meets the market. I've seen brilliant products languish in obscurity while mediocre ones soared because one team mastered their go-to-market strategy while the other treated it as an afterthought. After leading countless product launches across enterprise and consumer markets, I've learned that a methodical approach to launch planning is what separates consistent winners from occasional lucky shots.

This chapter builds upon everything we've covered about product development and prepares you for the critical transition from building to selling. Go-to-market (GTM) planning isn't just for new products. It applies equally to major features, expansions into new territories, or repositioning efforts.

Go-to-market planning sits at the intersection of product, marketing, sales, and customer success; making it the perfect arena for product leaders to demonstrate their cross-functional leadership skills. Too often, product teams and leaders throw their creation "over the wall" to marketing and sales, washing their hands of responsibility for market success. This is a grave mistake. As a product leader, you should be the primary architect of your go-to-market strategy, working closely with allied teams to ensure alignment and execution.

Many product managers excel at the "building" part but struggle with the "selling" part. It's often what separates a leader from a non-leader. This discomfort with commercialization can lead to an over-focus on product features rather than customer outcomes and business results. Understanding go-to-market planning helps you bridge this gap and communicate your product's value more effectively to both internal stakeholders and customers.

Successful go-to-market planning requires a holistic view of the customer journey, from initial awareness through purchase, adoption, and advocacy. By mapping each touchpoint and ensuring consistency across all interactions, you create a seamless experience that reduces friction in the buying process and accelerates adoption.

Channels & Distribution

The channels through which you deliver your product to market are as important as the product itself. Think of channels as the bridges connecting your product to your customers; if they're unstable, narrow, or in the wrong location, even the best product won't reach its intended audience.

Channel Strategy

Your channel strategy determines how customers will discover, evaluate, and purchase your product. The right mix of channels depends on your specific product, target audience, and business model.

Direct Sales

Direct sales channels give you the most control over the customer experience but require significant investment in sales teams and infrastructure. I've seen this approach work particularly well for complex enterprise products with high average contract values. Salesforce built its empire on a direct sales model, with account executives developing deep relationships with prospects and guiding them through lengthy sales cycles.

Partners / Indirect

Partner or indirect channels extend your reach by leveraging existing relationships and infrastructures. This approach multiplies your sales force without the overhead but sacrifices some control over messaging and customer experience. Microsoft's partner ecosystem has been central to their success, allowing them to scale globally while maintaining focus on their core technology.

Self-Service

Self-service channels enable customers to discover, evaluate, and purchase your product without human interaction. This approach scales efficiently but works best for products with straightforward value propositions and relatively low price points. Zoom's explosive growth during the pandemic was partly due to their frictionless self-service model that allowed users to sign up and start using the product within minutes.

Multi-Channel

Multi-channel strategies combine elements of the above approaches to serve different segments or stages of the customer journey. Enterprise customers might enter through direct sales, while SMB (small-to-medium business) customers use self-service, with partners handling specific verticals or regions. HubSpot exemplifies this approach, with a self-service model for small businesses, a partner network for agencies, and direct sales for enterprise customers.

Matching your channel strategy to your customer segments is critical. High-touch enterprise customers typically expect personal relationships and customized solutions, while small business or consumer customers often prefer self-service options with minimal friction. Your channel investments should align with the lifetime value of each customer segment.

Beyond initial selection, continuous monitoring of channel effectiveness is crucial — like checking your navigation system regularly. Markets evolve, customer preferences shift, and competitors introduce new routes. Without regular assessment, you might continue down an increasingly inefficient path while better alternatives emerge.

The most successful product leaders treat channel strategy as a living system rather than a one-time decision. They analyze performance metrics, gather customer feedback, and remain willing to pivot when the signs point to a better path forward.

Remember: the best channel isn't necessarily the most obvious or popular one; it's the one that best connects your unique product with your specific customers, creating the smoothest journey from problem to solution.

Example: CloudHarbor

CloudHarbor, a promising startup offering enterprise-grade cloud security solutions, initially chose what seemed the obvious path: a direct sales approach. After all, their complex product required technical explanation, customization, and relationship-building with IT security departments. Their team of skilled sales engineers conducted demos, negotiated contracts, and provided implementation support.

Six months in, CEO Shea Miller noticed concerning patterns in their metrics. While deals were closing, the sales cycle stretched uncomfortably long; averaging 97 days. Customer acquisition costs soared, and their small team struggled to scale beyond a handful of metropolitan markets.

During a quarterly review, CloudHarbor's product team revealed an unexpected insight: their most satisfied customers weren't the large enterprises they'd targeted, but mid-sized companies with limited internal security expertise. These customers valued CloudHarbor not just for its technology but for the guidance their sales engineers provided.

"We're not just selling software," Shea realized. "We're selling expertise these companies can't afford to hire internally."

This revelation prompted CloudHarbor to pivot to a partner-focused channel strategy, forming alliances with managed service providers (MSPs) already serving these mid-sized businesses. These partners bundled CloudHarbor's solution with their existing offerings, positioning it as a premium security enhancement.

The results transformed their business. The partner channel shortened sales cycles to 42 days and slashed customer acquisition costs by 61%. MSPs, already trusted advisors to their clients, more effectively communicated CloudHarbor's value proposition while managing implementation and support.

CloudHarbor's sales team shifted focus from direct selling to partner enablement; creating training materials, co-marketing programs, and incentive structures that aligned their interests with their channel partners.

Three years later, CloudHarbor maintains a small direct sales team for strategic accounts but generates 78% of revenue through partners; a channel alignment that better matches their resources, market position, and customer needs. Their willingness to pivot when data pointed to a better path ultimately led to sustainable growth rather than exhaustion in a misaligned channel.

Partnership Ecosystem

Strategic partnerships can dramatically accelerate your go-to-market efforts by providing instant credibility, expanded reach, and complementary capabilities. Building and managing these relationships requires deliberate planning and ongoing nurturing.

Please see the **Extensibility, Partnerships & Platform Integrations** chapter for in-depth coverage on this subject.

Technology partnerships integrate your product with complementary solutions to create more value for shared customers. These partnerships work best when there's natural alignment between user workflows. Slack's extensive integration marketplace has been central to their value proposition, allowing users to bring all their work into one collaborative environment.

Distribution partnerships leverage another company's established sales channels to reach new customers. These arrangements work particularly well when entering new markets where you lack presence or credibility. Adobe's partnership with Microsoft to distribute Creative Cloud through Office 365 channels opened new enterprise distribution avenues they couldn't have easily developed independently.

Co-marketing partnerships pool marketing resources to reach shared audiences with joint messaging. These arrangements amplify your voice and extend your budget. The partnership between GoPro and Red Bull exemplifies this approach, with each brand reinforcing the other's adventure-focused identity through co-created content and events.

Implementation partnerships ensure successful deployment and adoption of your product, particularly for complex solutions requiring specialized expertise. These partners become extensions of your customer success team. Salesforce's consulting partner network has been essential to their enterprise expansion, providing the industry-specific implementation resources that Salesforce couldn't scale internally.

When evaluating potential partners, look beyond short-term revenue opportunities to assess strategic alignment. The best partnerships are built on shared values, complementary capabilities, and mutual benefit. Both parties should contribute and gain value equitably, or the relationship will eventually falter.

Sales Enablement

Even the most experienced sales professionals can't effectively sell a product they don't understand or believe in. Sales enablement bridges the gap between product development and sales execution, ensuring your team has the knowledge, tools, and confidence to succeed.

Comprehensive product training forms the foundation of sales enablement. This includes not just feature knowledge but deep understanding of use cases, customer pain points, and competitive differentiation. Atlassian conducts regular "demo jams" where product managers present new features to sales teams in competitive scenarios, helping sales visualize real customer conversations.

Compelling sales collateral simplifies complex product information into practical tools that move deals forward. This includes presentations, data sheets, ROI calculators, and customer testimonials. ServiceNow maintains a centralized "deal hub" where sales reps can quickly access the latest approved materials for each product and customer segment.

Objection handling preparation arms your sales team with confident responses to common customer concerns and competitive threats. This typically includes battle cards, talk tracks, and regular competitive intelligence updates. Salesforce's competitive team publishes weekly intelligence briefings and runs scenario training to keep sales teams sharp against key competitors.

Sales playbooks codify your go-to-market approach into structured guidance for prospecting, discovery, demonstration, and closing. These playbooks should address different buyer personas, use cases, and deal complexities. IBM's sales approach includes detailed playbooks for each industry vertical, mapping product capabilities to specific industry challenges and regulatory requirements.

This probably goes without saying, but sales enablement is still super important for self-service sales channels. Anyone who's answering pre-sales or general inquiry questions needs to know all about the product's capabilities, just like traditional sales teams.

Effective sales enablement is continuous, not a one-time effort. Regular updates, on-going training, and feedback loops between sales and product teams ensure your go-to-market execution remains sharp as your product and market evolve. The best organizations treat sales enablement as a strategic function with dedicated resources, not an occasional task assigned to already-busy product managers.

Digital Presence

In today's digital-first buying environment, your online presence often creates the first impression for potential customers. A thoughtful digital strategy ensures this critical touchpoint accurately represents your value proposition and guides prospects toward conversion.

Your product website serves as the central hub of your digital presence. It should clearly communicate your value proposition, showcase key features, and provide clear paths to conversion based on buyer intent. Stripe's developer-focused website exemplifies this approach, with clean design, clear documentation, and interactive examples that allow developers to see the product in action immediately.

App marketplaces and directories have become crucial discovery channels for many products. Optimizing your presence in these environments requires understanding their unique algorithms and user behaviors. Zoom's App Marketplace listing success comes from clear categorization, comprehensive feature descriptions, and strategic use of screenshots and videos to showcase user experience.

Content marketing establishes thought leadership and builds trust with prospects before they're ready to buy. This approach is particularly effective for complex products with extended decision cycles. HubSpot's comprehensive resource center attracts millions of visitors monthly, creating a pipeline of marketing-qualified leads that enter their sales funnel.

Search optimization ensures your digital properties appear when prospects are actively seeking solutions to problems your product addresses. This requires ongoing keyword research, content optimization, and technical SEO work. Ahrefs has built their entire business around SEO tools, and they practice what they preach. Their own content dominates search results for competitive keywords in their industry.

Your digital presence doesn't exist in isolation. It should align with and support all other go-to-market channels, creating a consistent experience as customers move from digital to human interactions. The most effective digital strategies integrate with CRM and marketing automation systems to create seamless handoffs between digital engagement and sales follow-up.

Example: The Flower Box

The Flower Box, a small artisanal flower delivery startup, began with a simple model: locally-sourced, sustainable flowers delivered by bicycle messengers in Portland. For their first 18 months, they relied almost exclusively on word-of-mouth and local partnerships with wedding planners and event spaces.

While their business was steady, growth remained frustratingly slow. Their break-through moment came after a particularly disappointing Valentine's Day. Their courier team was fully staffed but orders totaled only 60% of their forecast. In their post-mortem analysis, the team discovered something startling: they were practically invisible online.

The founder spent a weekend setting up a basic website with professional photography of their arrangements and clear pricing. Within three weeks, their daily orders had tripled. Adding Instagram and Pinterest accounts showcasing their distinctive eco-friendly packaging and seasonal arrangements brought another wave of customers who had previously been unaware of their service.

The most revealing insight came when they implemented a simple "How did you hear about us?" question at checkout. Over 70% of new customers discovered them through Google searches for "sustainable flower delivery Portland" or similar terms; searches they had been invisible to previously.

Their most loyal customers had always loved their mission and products, but without a thoughtful digital presence, The Flower Box had been limiting their growth to only those customers who happened to hear about them through existing channels. What they had assumed was a product-market fit challenge was actually a fundamental distribution problem. They simply weren't visible where their ideal customers were looking.

Geographic Expansion Plan

Expanding into new geographic markets can unlock significant growth but requires careful planning and sequencing to avoid spreading resources too thin or stumbling over unforeseen obstacles.

Market prioritization should balance opportunity size with entry complexity. This analysis considers market size, competitive landscape, regulatory requirements, and cultural fit with your product. Shopify prioritized English-speaking markets initially before expanding into Europe and Asia, allowing them to refine their product and go-to-market approach before tackling more complex localization challenges.

Localization goes beyond simply translating your user interface and marketing materials. It requires adapting your product, messaging, and business practices to local norms and expectations. Airbnb adapted their payment methods for each market they entered, recognizing that payment preferences vary dramatically across regions.

Market entry models vary from light-touch digital presence to full local operations. The right approach depends on your product complexity, support requirements, and target customer profile. TransferWise (now Wise) initially entered new markets with a digital-only approach, establishing customer traction before investing in local teams and banking relationships.

Regulatory compliance requirements vary significantly across jurisdictions and industries. Overlooking these requirements can lead to costly delays or penalties. Stripe's methodical country expansion included detailed regulatory analysis and partnerships with local financial institutions to navigate complex payment regulations in each new market.

When expanding geographically, resist the temptation to customize your product extensively for each market. Instead, identify the minimum adaptations required for success while maintaining a globally consistent core experience. This balanced approach preserves operational efficiency while respecting local requirements.

Marketing & Demand Generation

Marketing transforms your product from an unknown entity to a recognized solution in the minds of potential customers. Effective demand generation creates a sustainable pipeline of qualified prospects who understand your value proposition and are predisposed to buy.

Marketing Mix Planning

The modern marketing landscape offers countless channels and tactics. The challenge isn't finding options; it's selecting the right mix to efficiently reach and influence your target audience. Let's talk about the 4 types of media exposure:

Paid Media

Paid media accelerates awareness and demand through purchased visibility. This category includes search advertising, social media promotion, display networks, and traditional advertising. DocuSign's aggressive paid search strategy captures prospects actively seeking electronic signature solutions, directing them to targeted landing pages with clear conversion paths.

Owned Media

Owned media builds your audience over time through channels you control directly. This includes your website, blog, email lists, and social profiles. HubSpot's comprehensive blog attracts millions of monthly visitors through SEO-optimized content addressing common marketing challenges, building an audience they can nurture toward conversion.

Earned Media

Earned media generates visibility through third-party coverage and word-of-mouth. This includes press coverage, analyst reports, customer reviews, and social sharing. Notion's explosive growth was significantly driven by user-generated content on social platforms, with enthusiasts sharing their custom workspaces and productivity systems.

Content Syndication

Content syndication extends the reach of your owned content through third-party distribution networks. This approach combines the control of owned media with the expanded reach of paid channels. Gartner's research reports are syndicated through multiple business publications, extending their influence while maintaining their thought leadership position.

Your marketing mix should evolve as your product matures and market conditions change. Early-stage products typically benefit from higher investment in awareness-building paid media, while established products can lean more heavily on owned and earned channels. Regular testing and measurement allow you to continuously optimize this mix based on actual performance data.

Example: Slumber Solutions

Slumber Solutions, a direct-to-consumer mattress company targeting environmentally-conscious millennials, crafted a marketing mix that leveraged multiple channels to build their brand.

For paid media, they allocated 40% of their budget to targeted Instagram and Pinterest ads showing their sustainable mattresses in aesthetically pleasing bedrooms, with additional investment in Google search ads capturing "eco-friendly mattress" keywords.

Their owned media strategy centered on their website featuring an interactive "Sleep Profile" quiz and a twice-monthly blog called "The Conscious Sleeper" covering topics from sleep science to sustainability practices. Their email program segmented subscribers based on quiz results, delivering personalized content about sleep improvements.

For earned media, they partnered with three mid-tier sleep and wellness influencers for honest reviews, resulting in significant social sharing. An unexpected boost came when their founder's appearance on a popular sustainability podcast generated a spike in website traffic from listeners interested in their manufacturing innovations.

When they discovered their strongest conversions came from the podcast appearances and Pinterest, they shifted budget from broader Instagram campaigns to more podcast sponsorships and expanded their Pinterest presence with "shoppable pins" featuring different bedroom styles.

Content Strategy Stages

Content marketing has evolved from a nice-to-have to an essential component of most B2B go-to-market strategies. A thoughtful content approach educates prospects, builds credibility, and nurtures relationships throughout the buyer's journey.

Awareness-Stage

Awareness content addresses broad pain points and challenges, helping prospects recognize and articulate their needs. This content typically avoids direct product promotion in favor of educational value. LinkedIn's Sophisticated Marketer's Guide series addresses high-level marketing challenges without pushing their platform, attracting prospects at the earliest stages of consideration.

Consideration-Stage

Consideration content helps prospects evaluate potential solutions and approaches to their recognized problems. This content can introduce your solution alongside alternatives, focusing on selection criteria and best practices. Salesforce's industry-specific solution guides help prospects understand how CRM might address their specific business challenges across different vendors.

Decision-Stage

Decision content directly supports the purchase process with product-specific information, comparisons, and validation. This includes case studies, ROI calculators, technical documentation, and implementation guidance. Zoom's security white papers directly address enterprise concerns about video conferencing privacy, removing a key objection in the final stages of purchase decisions.

Post-purchase-Stage

Post-purchase content supports successful implementation, adoption, and expansion. This content turns new customers into successful users and advocates. Slack's resource center provides robust onboarding guidance, use case examples, and best practices to help new customers realize value quickly.

The most effective content strategies maintain consistent themes and messages across all stages while adapting the format and depth to match the buyer's information needs at each point. This consistency creates a coherent narrative that guides prospects naturally from initial awareness to purchase decision.

I'm sure you've seen these staged content communications. You might check out an item on LinkedIn, and then you get an email saying how you'll benefit from buying it. Then, a week later, you get another email saying it's on sale. Next, you're on the website about to buy it, and more information is pushing you over the edge. After you buy it and a week later, you get another email about how powerful your purchase is. Isn't that effective?

Lead Generation Approach

Lead generation bridges marketing activities to sales opportunities through systematic identification and qualification of potential customers. A structured approach ensures marketing investments translate to revenue.

Inbound lead generation attracts prospects through valuable content and experiences, inviting them to self-identify their interest. This approach typically yields higher-quality leads with genuine interest in your solution. HubSpot built their entire business on inbound marketing principles, using educational content to attract marketing professionals who eventually become customers.

Outbound lead generation proactively identifies and engages potential customers who match your ideal customer profile. This approach provides more predictable volume but requires careful targeting to maintain efficiency. LinkedIn Sales Navigator enables precise targeting for outbound campaigns based on detailed company characteristics and professional criteria.

Lead qualification processes evaluate prospect fit and readiness before passing leads to sales teams. This critical step ensures sales resources focus on the highest-potential opportunities. Marketo's lead scoring model combines demographic attributes with behavioral signals to prioritize prospects showing genuine buying intent.

Lead nurturing programs maintain relationships with prospects who aren't yet ready to purchase, providing relevant content and touchpoints until the timing is right. These programs recognize that B2B buying cycles often extend over months or even years. Salesforce Pardot's automated nurture campaigns maintain engagement with early-stage prospects through personalized content journeys based on their specific interests and behaviors.

The most sophisticated lead generation approaches combine elements of both inbound and outbound, using data-driven insights to continuously refine targeting, messaging, and qualification criteria. Regular collaboration between marketing and sales teams ensures alignment on lead quality expectations and feedback loops for continuous improvement.

Launch Events & PR

Product launches represent concentrated opportunities to generate awareness, excitement, and momentum. Thoughtful planning ensures these moments create maximum impact and sustainable results.

Launch timing considerations include market readiness, competitive landscape, industry events, and seasonal factors. The ideal timing creates a window where your message can break through without excessive noise. Monday.com deliberately launched their product away from major tech events to ensure media attention wasn't divided among multiple announcements.

Announcement strategies determine how you'll introduce your product to the world. Options range from big splash events to progressive rollouts targeting specific segments. Slack famously launched with a personal email from founder Stewart Butterfield to technology journalists, creating an authentic introduction that sparked genuine interest rather than hype.

Media relations build relationships with journalists and outlets relevant to your market, ensuring your story reaches the right audiences. These relationships should begin well before launch day. Dropbox secured launch coverage in TechCrunch by providing founder Drew Houston's direct contact information, allowing for real-time responses to journalist questions.

Launch events bring your product to life through demonstrations, customer testimonials, and expert perspectives. These can be physical or virtual gatherings that create shared experiences around your product. Apple's product launch events have become the gold standard, creating anticipation and coverage that extends far beyond the tech press.

Post-launch momentum maintains visibility and interest after the initial announcement. This requires planned follow-up activities and content releases. Zoom maintained momentum after their initial public launch by releasing a steady stream of customer success stories demonstrating diverse use cases across industries.

Example: NutriPulse

NutriPulse, a startup developing a smart blender with nutritional analysis capabilities, crafted a launch strategy that captured media attention and generated customer excitement well beyond their initial budget.

The team made a strategic decision to launch during the January health and wellness season, but deliberately scheduled their announcement two weeks after CES to avoid being drowned out by major tech announcements. This timing sweet spot allowed health and tech journalists to give them focused attention during a traditionally quieter news period.

Instead of a traditional press release, NutriPulse's founder created personalized video demonstrations for twenty carefully selected journalists and influencers, showcasing how the blender analyzed the nutritional content of their favorite smoothie recipes. This unusual approach resulted in coverage from 15 of the 20 targeted outlets, with several journalists mentioning the personalized outreach as a refreshing change from standard PR pitches.

The centerpiece of their launch was a live-streamed "Blend-Off" event featuring three celebrity nutritionists creating custom smoothies while demonstrating the product's analysis features. The informal, engaging format generated over 50,000 concurrent viewers and created shareable moments that circulated on social media for weeks afterward.

Following the initial announcement, NutriPulse maintained momentum by releasing a series of user testimonial videos from their beta testing program, focusing on specific health goals their early users had achieved. These authentic stories provided credibility and demonstrated real-world applications that resonated with their target audience far more effectively than additional feature announcements would have.

Community Building

Communities transform transactional customer relationships into engaged ecosystems where users support each other, share best practices, and collectively increase the value of your product.

User forums provide spaces for customers to ask questions, share ideas, and solve problems together. These communities reduce support costs while building strong user relationships. Atlassian's community forums handle thousands of questions monthly, with experienced users often responding faster than official support teams.

Developer ecosystems extend your product through third-party integrations and extensions. These communities multiply your innovation capacity without expanding your team. Shopify's app ecosystem now includes over 6,000 extensions, allowing merchants to customize their stores for specific needs that Shopify itself couldn't address.

Customer advisory boards formalize relationships with your most strategic users, providing structured feedback channels and early influence on your roadmap. These groups offer invaluable perspective while creating strong advocates. Tableau's customer advisory board includes representatives from major enterprise customers who provide feedback on upcoming features and strategic direction.

Ambassador programs recognize and empower your most enthusiastic users to share their expertise and experiences. These programs create authentic advocacy that resonates more strongly than official marketing. Asana's ambassador program provides exclusive access, recognition, and resources to power users who demonstrate expertise and willingness to help others.

Community building requires genuine long-term commitment, not just marketing tactics. Successful communities are given real influence over product direction, recognition for their contributions, and authentic engagement from company leadership. The return on this investment comes through reduced churn, organic growth, and deeper market insights.

Post-Launch

The real work of product marketing begins after launch. The initial release provides real-world feedback that allows you to refine your approach based on actual market reception rather than assumptions.

Customer Success & Feedback Loops

Early customer experiences provide the ultimate reality check for your product and go-to-market assumptions. Establishing structured feedback channels ensures these insights drive ongoing improvement.

Onboarding experience monitoring identifies friction points in the critical first interactions with your product. These early moments disproportionately influence ultimate success or failure. Intercom implemented automated onboarding surveys at specific milestones, allowing them to identify and address friction points before they led to abandonment.

Voice of customer programs systematically collect and analyze customer feedback across touchpoints. These programs transform anecdotal feedback into actionable insights. Qualtrics built their experience management platform around structured voice of customer data, helping organizations identify and prioritize experience improvements.

Success metrics alignment ensures everyone agrees on how customer success will be measured. These metrics should focus on customer outcomes rather than just product usage. Salesforce's customer success teams track specific business metrics for each client, aligning product usage to tangible business results like increased sales efficiency or improved customer retention.

Closed-loop feedback processes communicate back to customers when their input influences product changes. These acknowledgments validate customer contributions and encourage continued engagement. Buffer's "Here's what we shipped" updates specifically reference customer requests that inspired new features, creating visible "you asked, we delivered" moments.

The most effective customer feedback programs balance quantitative metrics with qualitative insights, giving you both the what and the why behind customer behavior. Regular executive review of these insights ensures customer perspectives influence strategic decisions, not just tactical improvements.

Scaling Distribution Channels

Initial market traction reveals which acquisition channels deliver the best results for your specific product and audience. Scaling requires both optimizing these proven channels and thoughtfully exploring new avenues.

Channel economics analysis evaluates the true cost of customer acquisition across different sources, accounting for both direct expenses and conversion efficiency. This analysis often reveals surprising insights about your most profitable channels. Mailchimp discovered that their affiliate program delivered not only higher volume but also better retention rates than their paid search efforts, leading them to reallocate marketing budgets accordingly.

Multi-touch attribution models recognize that customers typically engage through multiple channels before converting. These models assign appropriate credit to each touchpoint, preventing misallocation of resources. Adobe's marketing attribution models helped them understand how their content marketing efforts supported conversion through other channels, even when content wasn't the final conversion point.

Channel diversification reduces dependence on any single acquisition source while reaching different audience segments. This approach provides resilience against unexpected channel changes. When Facebook ad costs spiked for direct-to-consumer brands, Harry's was less affected than competitors because they had already established strong podcast advertising and retail distribution channels.

Internationalization considerations affect channel selection and messaging as you expand geographically. Different markets often have dramatically different channel preferences and behaviors. Spotify's expansion into Japan required significant adaptation of their acquisition strategy, with greater emphasis on carrier partnerships and much less reliance on social media compared to their Western markets.

As you scale, resist the temptation to chase every possible channel. Instead, double down on channels that show strong unit economics while conducting controlled experiments to find new opportunities. The discipline to focus on proven channels while systematically testing alternatives separates sustainable growth from expensive churn.

Example: CircleTech

CircleTech, a productivity software startup, began with a simple Chrome extension that helped users organize their browser tabs. Their initial growth came exclusively through organic word-of-mouth and placement in the Chrome Web Store.

After analyzing their acquisition data, CircleTech discovered their customer acquisition cost through the Web Store was remarkably low, but growth had plateaued. Meanwhile, their early adopters were increasingly requesting a desktop version with expanded functionality.

The team made a strategic decision to double down on their most profitable channel while simultaneously testing new avenues. They optimized their Chrome Web Store listing with improved screenshots and feature descriptions, increasing conversion by 40% in that channel.

In parallel, CircleTech ran controlled experiments with three potential new channels: YouTube tutorial sponsorships, productivity podcasts, and a referral program. The podcast sponsorships delivered surprisingly strong unit economics. Each $2,000 sponsorship consistently generated $7,500 in annual recurring revenue.

When expanding internationally, CircleTech discovered their standard English-language marketing assets performed poorly in Japan, a market showing strong interest in their product. Rather than force-fitting their existing approach, they adapted by creating localized assets and partnering with local productivity influencers, eventually making Japan their second-largest market.

> The company's disciplined approach to channel economics analysis, experimentation, and market-specific adaptation allowed them to scale efficiently while avoiding the common startup trap of spreading resources too thinly across too many unproven channels.

Iteration & Product Roadmap Prioritization

The period immediately following launch provides rich insights that should influence your ongoing product development. Balancing quick improvements with strategic enhancements maintains momentum while building toward longer-term goals.

Quick wins identification focuses on high-impact, low-effort improvements that address immediate customer pain points. These rapid responses demonstrate responsiveness and build goodwill. Notion quickly addressed early user requests for improved mobile apps after their initial launch, recognizing this limitation was preventing adoption among otherwise enthusiastic users.

Roadmap reprioritization aligns upcoming development with actual market feedback rather than pre-launch assumptions. This adjustment ensures your resources focus on the most valuable opportunities. Discord significantly revised their roadmap after launch to emphasize reliability and voice quality over planned new features, responding to clear user priorities.

Communication cadence keeps customers informed about both immediate fixes and longer-term direction. This transparency builds trust and patience for more substantial improvements. Trello's public roadmap shows both recently completed items and upcoming priorities, giving users visibility into both recent progress and future direction.

Customer segmentation analysis informs feature prioritization by understanding which capabilities matter most to different user groups. This analysis helps balance the needs of various segments without trying to please everyone simultaneously. Slack's enterprise security features were prioritized based on clear feedback from their largest customers, even though these capabilities weren't priorities for their initial SMB user base.

The most effective post-launch prioritization maintains a balance between addressing immediate needs and pursuing strategic direction. Without this balance, products can become either unresponsive to customer feedback or directionless collections of reactive features without coherent vision.

Competitive Positioning Refinement

Market reception reveals which aspects of your positioning truly resonate and differentiate. Refining your messaging based on these insights ensures you emphasize your most compelling advantages.

Win/loss analysis examines the specific reasons customers choose your product or a competitive alternative. These insights often reveal surprising decision factors that weren't apparent pre-launch. Zoom discovered through win/loss analysis that reliability and simplicity were far more important differentiators than their advanced features, leading them to emphasize these aspects in their messaging.

Messaging effectiveness testing evaluates how well different value propositions resonate with target audiences. These tests identify which messages drive engagement, consideration, and conversion. Grammarly continuously tests different messaging approaches across their paid campaigns, finding that emphasizing productivity benefits outperformed both error correction and confidence-building messages.

Competitive response monitoring tracks how rivals adjust their positioning and offerings in response to your product. These movements often reveal your most threatening differentiation. When Figma launched with collaborative design capabilities, Adobe quickly began emphasizing their cloud collaboration features in Creative Cloud messaging, signaling they recognized this as a significant competitive threat.

User perception research assesses how actual customers describe your product's value and differentiation in their own words. This research often reveals messaging opportunities that marketing teams overlooked. Canva discovered through user interviews that many customers valued the confidence they gained from creating professional-looking designs more than the time savings, leading to a significant messaging shift.

Conclusion

Masterful go-to-market planning transforms great products into successful businesses. By thoughtfully addressing each element we've discussed — from channel strategy to competitive positioning — you'll greatly increase your chances of market success.

The best product leaders recognize that go-to-market planning isn't a one-time event but an ongoing discipline. Markets change, competitors respond, and customer needs evolve. The approaches outlined in this chapter provide not just launch playbooks but sustainable practices for continuous market adaptation.

Remember that your role as a product leader extends far beyond feature prioritization and development oversight. You are ultimately responsible for market success, which requires deep engagement with how your product reaches, acquires, and delights customers. By embracing this broader definition of product leadership, you position both your products and your career for sustained success.

Concepts in Action: Orchard

A startup named Orchard created an AI-powered app to help small farmers optimize their crop yields, reduce water usage, and connect directly with local buyers. Like seeds carefully planted in fertile soil, Orchard's founding team had developed a promising product, but they knew that even the most brilliant innovation could wither without the right go-to-market strategy.

The Channel Dilemma

Emma Chen, Orchard's product leader, stared at the whiteboard covered with potential distribution channels. Their initial assumption was that a direct sales approach made the most sense. After all, their product required careful explanation and farmers were notoriously hesitant about learning anything too technical.

"We can hire sales reps to visit farm shows and agricultural conferences," suggested their marketing director. "They'll demo the app and sign farmers up on the spot."

But after conducting field research, Emma discovered something unexpected. While individual farmers were indeed slow to adopt new technology, they had deep trust in their agricultural co-ops and extension offices. These organizations were already providing guidance on everything from seed selection to equipment purchases.

"We're not just selling software," Emma realized. "We're selling a transformative approach to farming that requires trust and education."

Like CloudHarbor in a previous example, Emma pivoted their channel strategy. Rather than building an expensive direct sales team, Orchard developed partnerships with agricultural cooperatives and extension services. These trusted advisors bundled Orchard's solution with their existing consultation services, positioning it as a premium tool that enhanced their expertise.

The results transformed their business. Their partner channel shortened sales cycles from 120 days to just 35 days and slashed customer acquisition costs by 70%. Co-ops, already trusted by farmers, communicated Orchard's value proposition more effectively while handling implementation and support.

The Marketing Mix

For their marketing approach, Orchard crafted a carefully balanced strategy that reminded Emma of the Slumber Solutions example from the chapter.

For paid media, they allocated budget to targeted ads in agricultural publications and sponsored content in farming podcasts, specifically focusing on sustainable farming keywords.

Their owned media centered on a practical blog called "Future Harvest" that shared both technical farming insights and customer success stories, while their email program segmented subscribers based on crop types and regional growing conditions.

For earned media, they partnered with respected agricultural scientists and sustainable farming advocates who tested their system and shared authentic results. An unexpected boost came when a popular farming YouTube channel did an extended feature on how one apple orchard increased yields by 32% while reducing water usage.

When they discovered their strongest conversions came from the podcast sponsorships and extension office partnerships, they reallocated resources to strengthen these channels.

Community Cultivation

Understanding that farmers often learn best from other farmers, Orchard invested heavily in community building. They created regional user groups where early adopters could share best practices and success stories with new users.

"These user groups became our most powerful sales tool," Emma explained to her investors. "When a skeptical farmer hears from their neighbor that our app helped them survive last summer's drought while increasing profits, it's more convincing than anything our marketing could ever say."

Like Atlassian's community forums mentioned earlier, Orchard's community became self-sustaining, with experienced users answering questions from newcomers faster than the official support team could respond.

Post-Launch Insights

After six months in market, Emma's team conducted a thorough analysis of user feedback and discovered something surprising. While their advanced AI forecasting features had been the primary focus of their development efforts, farmers were finding the most value in the simplified marketplace connecting them directly to local restaurants and grocers.

Similar to Discord's experience, Orchard significantly revised their roadmap to emphasize marketplace improvements over planned AI enhancements, responding to clear user priorities.

Through win/loss analysis, they discovered that competing apps offered more sophisticated analytics, but farmers chose Orchard because it addressed the entire journey from planting to selling; an insight that reshaped their messaging from "AI-powered farming" to "Farm-to-table made simple."

Scaling Success

As Orchard began to scale, they faced the channel economics decisions mentioned in earlier. Their analysis revealed that while agricultural co-ops remained their most efficient channel, the unit economics of their educational webinars were surprisingly strong; each $1,000 investment consistently generated $11,000 in annual recurring revenue.

When expanding internationally, they discovered their standard approach performed poorly in regions with different agricultural practices. Rather than force-fitting their existing strategy, they adapted by partnering with local agricultural institutions and customizing their app for regional growing conditions and market structures.

The Harvest

Three years later, Orchard maintained its partnership focus but had developed additional channels for different segments: a self-service option for tech-savvy younger farmers, an enterprise approach for large agricultural businesses, and their original partnership channel for traditional small-to-medium farms.

Their disciplined approach to go-to-market planning had transformed them from a promising seed of an idea into a flourishing business that was changing how small farms operated around the world.

"The moment of truth for any product isn't when the final line of code is written," Emma often reminded her team, echoing the chapter's opening line. "It's when it meets the market."

Execution Essentials

Channels & Distribution

- Match your channel strategy to your product complexity and customer value (direct sales for high-value, self-service for straightforward offerings).
- Align channels with customer segments (high-touch for enterprise, self-service for SMB).
- Treat channel strategy as a living system, not a one-time decision.
- Be willing to pivot when data reveals a misalignment between your channel and market.
- Build strategic partnerships that provide credibility, reach, and complementary capabilities.

Sales Enablement

- Invest in training that covers not just features but use cases, pain points, and competitive differentiation.
- Develop compelling sales collateral (presentations, ROI calculators, customer testimonials).
- Create structured sales playbooks for each stage of the selling process.
- Prepare thorough objection handling resources with battle cards and competitive intelligence.
- Make sales enablement a continuous process, not a one-time event.

Digital Presence

- Optimize your product website with clear value proposition and conversion paths.
- Establish thought leadership through targeted content marketing.
- Ensure strong search visibility for terms related to the problems you solve.
- Integrate your digital presence with CRM for seamless handoffs between digital engagement and sales.
- Create consistency across all digital touchpoints for a unified customer experience.

Marketing & Demand Generation

- Balance your marketing mix across paid, owned, earned, and syndicated media.
- Develop specific content strategies for each stage of the buyer's journey.

- Implement both inbound and outbound lead generation approaches.
- Use lead qualification processes to ensure sales teams focus on high-potential opportunities.
- Create lead nurturing programs to maintain relationships with prospects who aren't yet ready to buy.

Launch Events & PR

- Choose strategic launch timing that avoids major competitor announcements and industry noise.
- Build media relationships well before launch day.
- Create engaging launch events with demonstrations and customer testimonials.
- Plan post-launch activities to maintain momentum beyond the initial announcement.
- Adapt your announcement strategy based on your specific market and audience needs.

Community Building

- Establish user forums that enable customers to support each other.
- Create customer advisory boards for structured feedback and advocacy.
- Develop ambassador programs that empower enthusiastic users to share their expertise.
- Give your community genuine influence over product direction.
- Recognize and reward community contributions to build lasting engagement.

Post-Launch Strategies

- Implement closed-loop feedback processes to show customers their input matters.
- Analyze channel economics to identify your most efficient acquisition sources.
- Identify and implement quick wins based on early customer feedback.
- Reprioritize your roadmap based on actual market reception, not pre-launch assumptions.
- Conduct win/loss analysis to understand why customers choose your product or a competitor's.

*Good luck on your journey! Follow me on LinkedIn, find me at **thieros.com**, remember to put your customer in the center of everything you build, and smile while you still have teeth.*

Joey

Index

A

A/B Testing 284, 287
acquisition metrics 252
activation metrics 253
AI/ML Product Manager 73
APIs (Application Programming Interfaces) 305
authentication (with other platforms) 314

B

balancing workloads 96
benchmarking (features) 230
benefits (Employment) 30, 36, 42, 49, 56
bias mitigation 191
brand perception 229
business goals 201
Business Performance Metrics 209
  Cost-Benefit Analysis 209
  Determining Return on Investment 210
  Forecasting Revenue 209
  Resource Allocation Finances 210

C

career opportunities 98
Channel Dependency 256
Channel Strategy 324
Chief Product Officer 53
Circular Design Principles 182
coaching 165
  Asking the Right Questions 166
  clear outcomes 168
  problem-solving 167
  structured feedback 168
Cohort Analysis 286
collaboration (business) 212
  cross-functional 212
  Objectives & Key Results (OKRs) 213
  Portfolio Balancing 214
  Resource Capacity Planning 213

collaboration culture 137
Collaborative Workspaces 116
communication 132
Community Building 335
Community-Led Growth 258
company mission 202
compassion 101
Competitive Analysis 227
  Applying SWOT 232
  brand perception 229
  Competitive Intelligence Gathering 231
  competitor pricing 229
  Continuous Monitoring & Analysis 237
  customer reviews 234
  Digital Footprint Analysis 233
  Feature Release Monitoring 238
  Feature-to-Feature Benchmarking 230
  Gap Identification 235
  landscape shifts 239
  market positioning 228
  Market Share Tracking 238
  preemptive innovation 237
  pricing changes 239
  Roadmap Prioritization 235
  Strategic Response Planning 235
  value proposition mapping 228
  Win/Loss Analysis 232
Competitive Landscape Shifts 239
Competitive Positioning 203, 338
competitor pricing 229
complementary products 309
Compromise and Influence 120
Content Strategy Stages 331
content syndication 331
Cost-Benefit Analysis 209
cronyism 99
Cross-Functional Collaboration 212
cross-functional teams 129. See Also Leading Cross-Functional Teams
culture
  Cross-Functional Growth Teams 254
  Customer-Centric Culture 256
  experimentation 254

Customer Education 186
Customer Journey Mapping 205
Customer Journey Model, The 12
customer retention 250
Customer-Centric Culture 256
customization 308

D
data
    collection 283
    Collection Methods 284
    cross-functional 296
    integrity 283
    prioritization 297
    qualitative vs. quantitative 285
    quality 283
    Risk Assessment 294
    Sample Sizing 285
    storytelling 295
    visualization 294
data analysis
    A/B Testing 287
    cohort 286
    Funnel Analytics 287
    Predictive Modeling 288
data collection 283
Data Infrastructure 254
Data Integrity 283
data prioritization 297
Data Product Manager 75
data quality 283
Data Visualization 294
Decision Trees 293
Deferred Monetization Delay, The 277
Define, Relate, Communicate 109
Defining Stakeholders 109
defining success 122
Deliberate Retention 250
dependencies (platform) 313
Deployment Planning 217
Developing Future Leaders 173
development velocity 216
digital footprint 233
Digital Presence 328
direct sales 324

Director of Product Management 39
Discount Addiction, The 276
distribution channels, scaling 336
documentation 306

E
earned media 330
Ecosystem Thinking 258
emotional intelligence 97
Energy Efficiency 186
Enterprise Licensing 271
ethical leadership 181
    Accountability 188
    Bias Mitigation 191
    education (continuous) 192
    Inclusive Design 189
    privacy 190
    Purpose-Driven Metrics 192
    testing methods 191
    Transparency 188
    User-Centered Decision Making 187
evaluating partnerships 310
Execution & Operation 216
    deployments 217
    development velocity 216
    Operational Scalability 218
    Quality Assurance Standards 216
    technical debt 217
expansion 251
    metrics 253
Experimentation Culture 254
extensibility 304
    APIs (Application Programming Interfaces) 305
    customization 308
    documentation 306
    modular architectures 304
    plugins 307

F
Feature Bloat 257
feature planning 121
Feature-Millstone Effect, The 276
feedback loops 336
Forecasting Revenue 209
Freemium Pricing 270

Functional Model, The 13
Funnel Analytics 287

G

Gap Identification 235
geographic expansion 329
Go-To-Market Planning 323
    Channel Strategy 324
    Community Building 335
    competitive positioning 338
    Content Strategy Stages 331
    content syndication 331
    Digital Presence 328
    direct sales 324
    distribution channels, scaling 336
    earned media 330
    feedback loops 336
    geographic expansion 329
    launch events 333
    lead generation 333
    owned media 330
    paid media 330
    partner/indirect sales channel 324
    Partnership Ecosystem 326
    public relations 333
    roadmap prioritization 338
    Sales Enablement 327
    self-service sales channel 324
goals 130
governance (partnerships) 311
green materials 184
    selection 184
Group Product Manager 33
Growth
    Channel Dependency 256
    future of, the 258
    holistic customer experience 257
    Premature Scaling 256
    teams 254
Growth Product Manager 69
growth-oriented organization 254

H

Hidden Cost Oversight, The 277
holistic customer experience 257
hybrid teams 149

I

Inclusive Design 189
Innovation Balance 186

K

keeping promises 89

L

launch events 333
lead generation 333
leadership legacy 102
leadership roles
    Chief Product Officer 53
    Director of Product Management 39
    Group Product Manager 33
    misalignment 22
    proper title structure 23
    Senior Product Manager 27
    Vice President of Product 45
    why titles matter 21
Leading Cross-Functional Teams 129
    adjusting goals 132
    Celebrating Successes 139
    effective meetings 135
    goals 130
    key performance indicators (KPIs) 131
    Meetings and Communication 132
    objectives 130
    open dialog 134
    psychological safety 139
Leading vs. Lagging Indicators 289
Learning to Say No 120
Lifecycle Analysis 182

M

Market & Customer Alignment 204
    Customer Journey Mapping 205
    Market Timing Analysis 205
    Segmenting Target Users 207
    solutions 205
    voice of customer 207
market differentiation 228
market positioning 228
Market Share Tracking 238
Market Timing Analysis 205
mastering data 283

Meaningful Activation 249
meetings (minimizing) 154
Meetings & Communication 132
Mentoring
    accountability 171
    Balancing Support 171
    Career Development 169
    commitment 170
    Creating a Safe Space 169
    unwritten rules 169
merit 99
metrics 289
    Business Performance Metrics 209
    engagement 290
    for sustainability 183
    leading vs. lagging indicators 289
    purpose-driven 192
    retention 291
modular architectures 304
monetization 267
    Communication Strategy 273
    expansion 274
    Initial Monetization 274
    modeling 269
        Enterprise Licensing 271
        Freemium Models 270
        Marketplace/Transaction Models 270
        Subscription Models 269
        Usage-Based Pricing 270
    pitfalls
        Deferred Monetization Delay, The 277
        Discount Addiction, The 276
        Feature-Millstone Effect, The 276
        Hidden Cost Oversight, The 277
    Results Visualization 268
    strategy 271
    success metrics 268
    timing 271
    Value Alignment 275
    Value Anchoring 268
    value perception 267

N

Newsletters 115
North Star 202
    metric 289

O

Objectives & Key Results (OKRs) 213

on-site teams 149
Operational Scalability 218
Outcome-Based Model, The 13
owned media 330

P

paid media 330
partner/indirect sales channel 324
partnership lifecycles 311
partnerships 303
Platform Integration 313
    authentication 314
    data synchronization 314
    dependencies 313
    fallback mechanisms 315
    Return on Investment (ROI) 316
Platform Product Manager 71
plugins 307
politics 95
Portfolio Balancing 214
Post-Launch 335
Predictive Modeling 288
preemptive innovation 237
Premature Scaling 256
pricing 269
    communication 273
    sales channels 272
    strategy 211
    testing 272
privacy-first 258
problem-solution fit 205
product alignment 201
    Aligning to Company Mission 202
    Competitive Positioning 203
    North Star 202
    roadmap planning 203
    Strategic Vision 201
    value proposition 203
product architecture 12
    choosing 14
    Customer Journey Model, The 12
    Functional Model, The 13
    Outcome-Based Model, The 13
    testing 15
Product Extensibility 304
product-centric organizations 3

cross-functional collaboration 5
Customer Obsession 4
decision-making 4
defining and adopting 3
executive courage 9
failure 7
metrics 6
path to 9
Product-Led Growth 247
psychological safety 139
public relations 333
Pulse Surveys 115

Q
Quality Assurance Standards 216

R
raw data 136
recharging 101
recognition 100
Regulatory Compliance 184
remote teams 149
accountability 157
discipline 156
inclusiveness 150
resource allocation 210
Resource Capacity Planning 213
retention (customer) 250
metrics 253
Return on Investment (ROI) 210
from platform integration 316
Risk Assessment (data) 294
roadmap prioritization 338
roadmaps 203
long-term planning 203
prioritization 235
roles. See leadership roles

S
salaries (for various roles) 30, 36, 42, 49, 56
sales channels 272
Sales Enablement 327
scaling 256
Segmenting Target Users 207

self-service sales channel 324
Senior Product Manager 27
Senior Technical Product Manager 67
servant leadership
behaviors 173
building consensus 172
Developing Future Leaders 173
empowering 173
measuring your success 174
promoting your team 172
Removing Obstacles 172
Show-and-Tells 116
Specialized Product Roles 65
AI/ML Product Manager 73
Data Product Manager 75
Growth Product Manager 69
Platform Product Manager 71
Senior Technical Product Manager 67
Technical Product Manager (TPM) 67
Stakeholder Management 109
Collaborative Workspaces 116
communication 115
Compromise and Influence 120
Define, Relate, Communicate 109
Defining Stakeholders 109
empathy 112
expectation management 214
feature planning 121
group engagement 113
Hidden Influencers 110
inequality 112
long vs. short-term 112
Newsletters 115
Pulse Surveys 115
relationships 111
saying no 120
segments 110, 116
sensitive information 122
Show-and-Tells 116
sign-on / sign-off 112, 117
stakeholder grid 111
Stakeholder Mapping 110
success, defining 122
team involvement 113
tools 122
transparency 122
Stakeholder Mapping 110
Stakeholder Segments 110, 116
Stakeholders. See Stakeholder

Management
Strategic Expansion 251
   North Star 202
Strategic Partnerships
   complementary products 309
   evaluating partnerships 310
   governance 311
   mutual value 310
   partnership lifecycles 311
Strategic Planning 235
Strategic Vision 201
strategy 227
structured feedback 168
Subscription Models (monetization) 269
success metrics 268
surveys 284
sustainability 181
   balancing priorities 181
   Circular Design Principles 182
   Customer Education 186
   Energy Efficiency 186
   Innovation Balance 186
   Lifecycle Analysis 182
   metrics 183
   Regulatory Compliance 184
   Stakeholder Alignment 183
sustainable growth 248
SWOT 232
   example 233

T

taking breaks (work-life balance) 97
team development
   assessments and expectations 90
   authority, evolving 88
   Balancing Workloads 96
   buddy system 93
   compassion for members 101
   coordination 85
   cronyism 99
   culture of trust 89
   documentation 94
   emotional intelligence 97
   escalations 87
   feedback 94
   hierarchy 83, 87
   leadership legacy 102

merit 99
micromanagement 84, 87
Onboarding for Future Success 92
onboarding investment 94
one-on-ones 95
opportunities 98
politics 95
recharging 101
recognition of members 100
setting clear expectations 94
supporting 85
taking breaks 97
Trust Culture Reward, The 91
uncomfortable conversations 90
work-life balance 97
technical debt 217
Technical Product Manager (TPM) 67
testing methods, ethical 191
Thoughtful Acquisition 249
transaction pricing 270
transparency
   mistakes 122
   supply chain 183

U

unit economics 257
Usage-Based Pricing 270
User-Centered Decision Making 187

V

Value Anchoring 268
value perception 267
value proposition 203
   mapping 228
Vice President of Product 45
voice of customer 207

W

Win/Loss Analysis 232
work-life balance 97

www.ingramcontent.com/pod-product-compliance
Lightning Source LLC
Chambersburg PA
CBHW081801200326
41597CB00023B/4107